THE WEATHER MAKERS

D1531339

THE
WEATHER
MAKERS

THE HISTORY AND FUTURE IMPACT OF CLIMATE CHANGE

TIM
FLANNERY

ALLEN LANE

an imprint of

PENGUIN BOOKS

ALLEN LANE

Published by the Penguin Group
Penguin Books Ltd, 80 Strand, London WC2R ORL, England
Penguin Group (USA) Inc., 375 Hudson Street, New York, New York 10014, USA
Penguin Group (Canada), 90 Eglinton Avenue East, Suite 700, Toronto, Ontario, Canada M4P 2Y3
(a division of Pearson Penguin Canada Inc.)
Penguin Ireland, 25 St Stephen's Green, Dublin 2, Ireland (a division of Penguin Books Ltd)
Penguin Group (Australia), 250 Camberwell Road,
Camberwell, Victoria 3124, Australia (a division of Pearson Australia Group Pty Ltd)
Penguin Books India Pvt Ltd, 11 Community Centre,
Panchsheel Park, New Delhi – 110 017, India
Penguin Group (NZ), cnr Airborne and Rosedale Roads, Albany,
Auckland 1310, New Zealand (a division of Pearson New Zealand Ltd)
Penguin Books (South Africa) (Pty) Ltd, 24 Sturdee Avenue,
Rosebank, Johannesburg 2196, South Africa

Penguin Books Ltd, Registered Offices: 80 Strand, London WC2R ORL, England

www.penguin.com

First published in Australia by The Text Publishing Company 2005
First published in Great Britain by Allen Lane 2006

3

Copyright © Tim Flannery, 2005

Printed in Great Britain by Clays Ltd, St Ives plc

A CIP catalogue record for this book is available from the British Library

HARDBACK
ISBN-13: 978-0-713-99921-1
ISBN-10: 0-713-99921-7

TRADE PAPERBACK
ISBN-13: 978-0-713-99930-3
ISBN-10: 0-713-99930-6

This book is produced on Munken Premium (cream) paper which is an
FSC certified paper supplied by Arctic Paper Munkedals AB, Sweden.

*To David and Emma, Tim and Nick, Noriko and Naomi,
Puffin and Galen, Will, Alice, Julia and Anna, and of course Kris,
with love and hope; and to all of their generation who will
have to live with the consequences of our decisions.*

CONTENTS

PART I

GAIA'S TOOLS

A great-aunt's musings on the atmosphere. Wallace's astonishing aerial ocean. Lovelock's heresy: the data is thin and yet it lives. Ice crossing the Line—until plankton tune the thermostat. The importance of albedo. Creating coal—another Gaian fine-tuning?

The atmosphere's four layers and the mystery of why, though closer to the Sun, mountains are cold. The window in the brick wall of gases. A midsummer night's nightmare in New York. Earth's inspiration and indispensable interconnector. Pollution changes her nature and moods. Watching from Mauna Loa as the world draws breath.

Early disbelief at the power of CO_2. A mighty tough carbon budget. The thirty gases that heat the world. Methane: swamps, farts and belches. The CFCs—Frankenstein creations of human ingenuity. Where do all the gigatonnes go? Earth's carbon lungs, carbon stores, carbon kidney and carbon Gaia. Lessons from a can of soda. The misleading Mississippi.

PART 3

THE SCIENCE OF PREDICTION

industry. Both define our commitment, but George Dubya considerably extends it. The threshold of extreme danger—400 or 1200 parts per million; or have we already crossed it?

PART 4

PEOPLE IN GREENHOUSES

and dead birds. The solar thermal-wind challenge. And then
there is light—wonders of the photovoltaic cell.
How long the paybacks?

FOREWORD

Over the past four years I have had the pleasure of working with Tim Flannery as part of the Wentworth Group of Concerned Scientists. This gathering of eminent scientists was established to provide workable solutions to the key Australian environmental issues of land and water management. It elevated these issues to the national agenda, and helped achieve unprecedented environmental outcomes. But all our work and that of conservationists the world over could be rendered redundant as a result of the impacts of climate change.

We now stand at a threshold and face alternative futures: one too awful to contemplate, and one where we can continue to thrive and prosper, but within the ecological limits of the natural world we inhabit. *The Weather Makers*, as the title suggests, makes it clear that we have just enough time to choose which of these futures we want.

This book also makes it clear that the consequences of climate change are so profound and far-reaching that they will affect every aspect of our lives, our economies and societies. Quite simply, climate change is a threat to civilisation as we know it. This is an issue for everyone, not merely for a small band of environmental activists or an elite of international policy-makers: governments and industries in particular will need to adopt a courageous and vital leadership role. The solutions, however, do not lie solely in the domain of policy or technical fixes. If we are to win the war on climate change we must all be part of the fight.

The Weather Makers will challenge you to think about the changes you could make in your own life. I don't believe that anyone can read this book and not be moved to act. We do still have time to avert disaster, but there is not a moment to lose.

Robert Purves
President, WWF Australia
July 2005

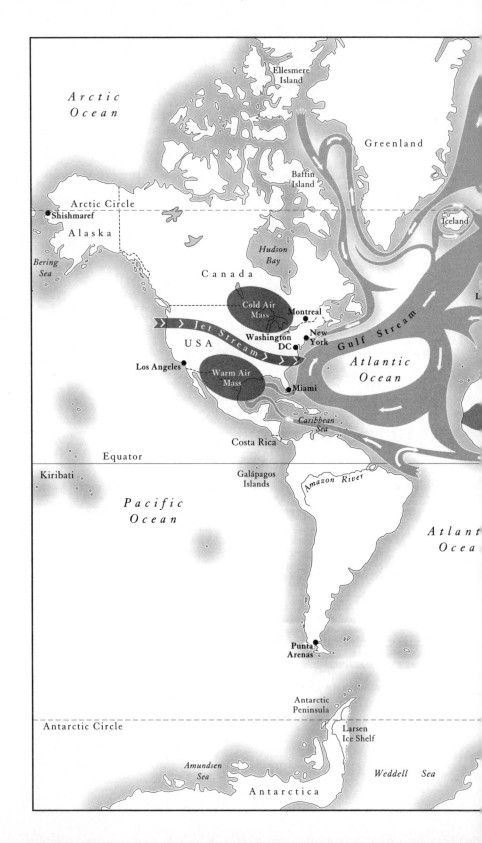

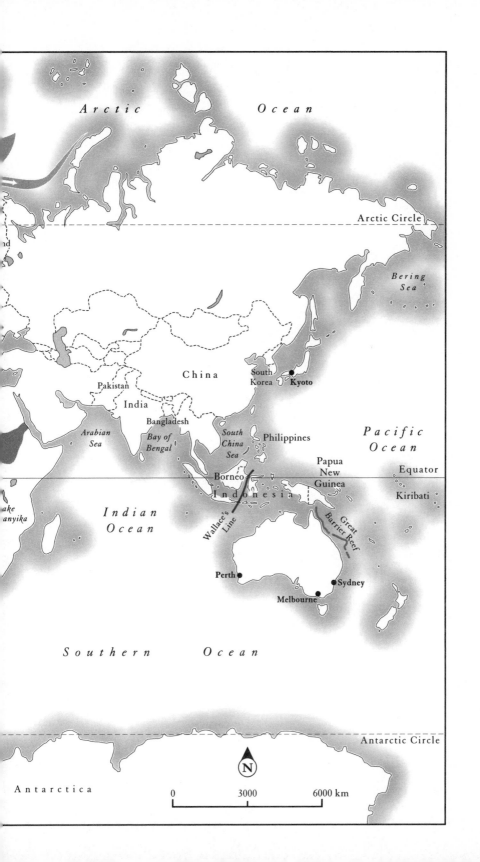

Arctic Ocean

Arctic Circle

Bering Sea

China

South Korea

Kyoto

Pakistan

India

Bangladesh

Arabian Sea

Bay of Bengal

South China Sea

Philippines

Pacific Ocean

Papua New Guinea

Equator

Kiribati

Borneo

Indonesia

Indian Ocean

Wallace's Line

Great Barrier Reef

Lake Tanganyika

Perth

Melbourne

Sydney

Southern Ocean

Antarctic Circle

Antarctica

N

0   3000   6000 km

THE SLOW AWAKENING

In 1981, when I was in my mid-twenties, I climbed Mt Albert Edward, one of the highest peaks on the verdant island of New Guinea. Although only 120 kilometres from Papua New Guinea's national capital, Port Moresby, the region around Mt Albert Edward is so rugged that the last significant biological work conducted there was by an expedition from the American Museum of Natural History in the early 1930s.

The bronzed grasslands were a stark contrast to the green jungle all around, and among the tussocks grew groves of tree-ferns, whose lacy fronds waved above my head. Wallaby tracks threaded from the forest edge to the herbfields that flourished in damp hollows, and the scratchings and burrows of metre-long rats and the traces where long-beaked echidnas had probed for worms were everywhere. Many of these creatures, I later discovered, were unique to such alpine regions.

Downslope, the tussock grassland ended abruptly at a stunted, mossy forest. A single step could carry you from sunshine into the dank gloom, where the pencil-thin saplings on the margin were so festooned with moss, lichens and filmy ferns that they ballooned to the diameter of my waist. In the leaf litter on the forest floor I was surprised to find the trunks of dead tree-ferns. Tree-ferns grew only in the grassland, so here was clear evidence that the forest was colonising the slope from below. Judging from the distribution of the tree-fern trunks, it had swallowed at least thirty metres of grassland in less time than it takes for a tree-fern to rot on the damp forest floor—a decade or two at most.

Why was the forest expanding? As I pondered the mouldering trunks I remembered reading that New Guinea's glaciers were melting. Had the temperature on Mt Albert Edward warmed enough to permit trees to grow where previously only grasses could take root? And, if so, was this evidence of climate change? My doctoral studies were in palaeontology, so I knew how important changes in climate have been in determining the fate of species. But this was the first evidence I'd seen that it might affect Earth during my lifetime. The experience left me troubled; I knew there was something wrong, but not quite what it was.

Despite the good position I was in to understand the significance of these observations, I soon forgot about them. This was partly because, as I studied the various ancient ecosystems that our generation has inherited, seemingly bigger and more urgent issues demanded my attention. And some of the crises did seem dire: the rainforests that I was studying were being felled for timber and to make agricultural land, and the larger animal species living there were being hunted to extinction. In my own country of Australia rising salt was threatening to destroy the most fertile soils, while overgrazing, degradation of waterways and the logging of forests all threatened precious ecosystems and biodiversity. To me these were the truly pressing issues.

Whether we are crossing the road or paying the bills, it is the big, fast-moving things that command our attention. But seemingly large issues sometimes turn out to be a sideshow. The Y2K bug is one such example. Around the globe many governments and companies spent billions to prepare themselves against the threat, while others spent nothing; and yet 1999 gave way to 2000 with barely a hiccough, let alone an apocalypse. A sceptical eye is our greatest asset in dealing with this type of 'problem'. And deep scepticism has a particularly important role to play in science, for a theory is only valid for as long as it has not been disproved. Scientists are in fact trained sceptics, and this eternal questioning of their own and others' work may give the impression that you can always find an expert who will champion any conceivable view.

While such scepticism is the lifeblood of science, it can have drawbacks when society is called on to combat real dangers. For decades both the tobacco and asbestos industries found scientists prepared publicly to be doubtful about discoveries linking their products with cancer. A non-specialist cannot know whether the view being presented is fringe or mainstream thinking, and so we may come to believe that there is a real division in the scientific community on these matters. In the case of asbestos and tobacco, the situation was made worse because cancers often appear years after exposure to carcinogenic products, and no one can say for certain just who, among the many exposed, will be struck down. By creating doubt about the link between their products and cancer, the tobacco and asbestos companies enjoyed decades of fat profits, while millions of people met terrible deaths.

And many people have reacted with rightful caution to news about climate change. After all, we have in the past got things badly wrong.

In the 1972 publication *The Limits to Growth*, the Club of Rome told us the world was running out of resources and predicted catastrophe within decades. In an era of excessive consumption this imagined drought of raw materials gripped the public imagination, even though no one knew with any degree of certainty what volume of resources lay hidden in the earth. Subsequent geological exploration has revealed just how wide of the mark our estimates of mineral resources were back then, and even today no one can accurately predict the volume of oil, gold and other materials beneath our feet.

The climate change issue is different. It results from air pollution, and the size of our atmosphere and the volume of pollutants that we are pouring into it are known with great precision. The debate now, and the story I want to explore here, concerns the impacts of some of those pollutants (known as greenhouse gases) on all life on Earth.

Is climate change a terrible threat or a beat-up? A bang or a whimper? Perhaps it's something in between—an issue that humanity must eventually face, but not yet. The world's media abound with evidence to support any of these views. Yet perusing that same media makes one thing clear: climate change is difficult for people to evaluate dispassionately because it entails deep political and industrial implications, and because it arises from the core processes of our civilisation's success. This means that, as we seek to address this problem, winners and losers will be created. The stakes are high, and this has led to a proliferation of misleading stories as special interest groups argue their case.

What's more, climate change is a breaking story. Just over thirty years ago the experts were at loggerheads about whether Earth was warming or cooling—unable to decide whether an ice-house or a greenhouse future was on the way. By 1975, however, the first sophisticated computer models were suggesting that a doubling of carbon dioxide (CO_2) in the atmosphere would lead to an increase in global temperature of around 3°C. Still, concern among both scientists and the community was not significant. There was even a period of optimism when some researchers believed that extra CO_2 in the atmosphere would fertilise the world's croplands and produce a bonanza for farmers.

But by 1988 climate scientists had become sufficiently worried about CO_2 to establish a panel, staffed with the world's leading experts, to report twice each decade on the issue. Their third report, issued in 2001, sounded a note of sober alarm—yet many governments and industry leaders were slow to take an interest. Because concern about climate change is so new, and the issue is so multi-disciplinary, there are few true experts in the field, and even fewer who can articulate what the problem might mean to the general public and what we should do about it.

For years I had resisted the impulse to devote research time to climate change. I was busy with other things, and I wanted to wait and see, hoping an issue so big would sort itself out. Perhaps it would be centuries before we would need to think intensively about it. But by 2001, articles in scientific journals indicated that the world's alpine environments were under severe threat. As I read them I remembered those rotting tree-fern trunks in Mt Albert Edward's forest, and I knew that I had to learn more.

This meant teaching myself about greenhouse gases, the structure of our atmosphere, and how the industrialised world powers its engines of growth.

For the last 10,000 years Earth's thermostat has been set to an average surface temperature of around 14°C. On the whole this has suited our species splendidly, and we have been able to organise ourselves in a most impressive manner—planting crops, domesticating animals and building cities. Finally, over the past century, we have created a truly global civilisation. Given that in all of Earth history the only other creatures able to organise themselves on a similar scale are ants, bees and termites—which are tiny in comparison and have concomitantly small resource requirements—this is quite an achievement.

Earth's thermostat is a complex and delicate mechanism, at the heart of which lies carbon dioxide, a colourless and odourless gas. CO_2 plays a critical role in maintaining the balance necessary to all life. It is also a waste product of the fossil fuels that almost every person on the planet uses for heat, transport and other energy requirements. On dead planets such as Venus and Mars, CO_2 makes up most of the atmosphere, and it would do so here if living things and Earth's processes did not keep it within bounds. Our planet's rocks and waters are packed with carbon itching to get airborne and oxidised. As it is, CO_2 makes up around three parts per 10,000 in Earth's atmosphere. It's a modest amount, yet it has a disproportionate influence on the planet's temperature. Because we create CO_2 every time we drive a car, cook a meal or turn on a light, and because the gas lasts around a century in the atmosphere, the proportion of CO_2 in the air we breathe is rapidly increasing.

The institutions at the forefront of climate change research are situated half a world away from my home in Adelaide, so for a time I flew frequently across the globe. One night when en route from Singapore to London, as we crossed the great Eurasian landmass, I looked out of the cabin window at a city illuminated below. Its network of lights stretched from horizon to horizon, and the lights burned so bright—with so much energy—as to alarm me. From a height of 10,000 metres the atmosphere seemed so thin and fragile—the breathable part of it lay 5000 metres below our aircraft. I asked the airline steward where we were. She gave me the name of a city I didn't know. With a jolt I realised that the world is full of such cities, whose fossil-fuel-driven lights cause our planet to blaze into the night sky.

By late 2004, my interest had turned to anxiety. The world's leading science journals were full of reports that glaciers were melting ten times faster than previously thought, that atmospheric greenhouse gases had reached levels not seen for millions of years, and that species were vanishing as a result of climate change. There were also reports of extreme weather events, long-term droughts and rising sea levels.

For months I tried to fault the new research findings, and discussed them at length with friends and colleagues. Only a few seemed aware of the great changes under way in our atmosphere. And some people I loved and respected continued doing things—such as buying large cars and air conditioners—which I now suspected to be very bad indeed.

By the end of the year, however, glimmers of hope were beginning to emerge, with almost every head of government in the developed world alive to the issue. But we cannot wait for the issue to be solved for us. The most important thing to realise is that we can all make a difference and help combat climate change at almost no cost to our lifestyle. And in this, climate change is very different from other environmental issues such as biodiversity loss or the ozone hole.

The best evidence indicates that we need to reduce our CO_2 emissions by 70 per cent by 2050. If you own a four-wheel-drive and replace it with a hybrid fuel car, you can achieve a cut of that magnitude in a day rather than half a century. If your electricity provider offers a green option, for the cost of a daily cup of coffee you will be able to make equally major cuts in your household emissions. And if you vote for a politician who has a deep commitment to reducing CO_2 emissions, you might change the world. If you alone can achieve so much, so too can every individual and, in time, industry and government on Earth.

The transition to a carbon-free economy is eminently achievable because we have all the technology we need to do so. It is only a lack of understanding and the pessimism and confusion generated by special interest groups that is stopping us from going forward.

One thing that I hear again and again as I discuss climate change with friends, family and colleagues, is that it is something that may affect humanity in decades to come, but is no immediate threat to us. I'm far from certain that this is true, and I'm not even sure it is relevant. If serious change or the effects of serious change are decades away, that is just a long tomorrow. Whenever my family gathers for a special event, the true scale of climate change is never far from my mind. My mother, who was born during the Great Depression—when motor vehicles and electric lights were still novelties—positively glows in the company of her grandchildren, some of whom are not yet ten. To see them together is to see a chain of the deepest love that spans 150 years, for those grand-children will not reach my mother's present age until late this century. To me, to her, and to their parents, their welfare is every bit as important as our own. On a broader scale, 70 per cent of all people alive today will still be alive in 2050, so climate change affects almost every family on this planet.

A final issue that looms large in discussions is the one of certainty. Four nations are yet to sign the Kyoto Protocol limiting CO_2 emissions: the USA, Australia, Monaco and Liechtenstein. President George W. Bush has said he wants 'more certainty' before he acts on climate change; yet science is about hypotheses, not truths, and no one can absolutely know the future. But this does not stop us making forecasts and modifying our behaviour accordingly. If, for example, we wait to see if an ailment is indeed fatal, we will do nothing until we are dead. Instead we take medication or whatever else the doctor dispenses, despite the fact that we may survive regardless. And when it comes to more mundane matters, uncertainty hardly deters us: we spend large sums on our children's education with no guarantee of a good outcome, and we buy shares with no promise of a return. Excepting death and taxes, certainty simply does not exist in our world and yet we often manage our lives in the most efficient manner. I cannot see why our response to climate change should be any different.

One of the biggest obstacles to making a start on climate change is that it has become a cliché before it has even been understood. What we need now is good information and careful thinking, because in the years to come this issue will dwarf all the others combined. It will become the *only* issue. We need to re-examine it in a truly sceptical spirit—to see how big it is and how fast it's moving—so that we can prioritise our efforts and resources in ways that matter.

What follows is my best effort, based on the work of thousands of colleagues, to outline the history of climate change, how it will unfold over the next century, and what we can do about it. With great scientific advances being made every month, this book is necessarily incomplete. That should not, however, be used as an excuse for inaction. We know enough to act wisely.

I

GAIA'S TOOLS

ONE

GAIA

> There must be an intricate security system to ensure that exotic outlaw species do not evolve into rampantly criminal syndicates…
>
> When a species…produces a poisonous substance, it may well kill itself. If, however, the poison is more deadly to its competitors it may manage to survive and in time both adapt to its own toxicity and produce even more lethal forms of pollutant.
>
> James Lovelock, *Gaia*, 1979.

Until a black mood takes her and she rages about our heads, most of us are unaware of our atmosphere. The 'atmosphere': what a dull name for such a wondrous thing. And it's hardly specific. I remember, as child, my great-aunt sitting with my mother at our kitchen table, a cup of tea in hand, saying meaningfully, 'You could have cut the atmosphere with a knife.' If we took the same linguistic approach to things maritime we would use the catch-all word 'water' to replace 'sea' and 'ocean', leaving us with no way to indicate whether we meant a glassful or a half a planet's worth of hydrogen oxide, as H_2O is properly known.

It was Alfred Russel Wallace, co-founder with Charles Darwin of the theory of evolution by natural selection, who came up with the phrase 'The Great Aerial Ocean' to describe the atmosphere. It's a far better name, because it conjures in the mind's eye the currents, eddies and layers that create the weather far above our heads, and which is all that stands between us and the vastness of space. Wallace's phrase was born of a romantic era of scientific discovery when both amateurs and professionals were making significant contributions towards understanding why cyclones rage in certain regions of the globe, and how 'carbonic acid', as carbon dioxide was sometimes described, affects the distributions of plants and animals.

Reading such work, you get the sense that their discoveries caused as much excitement as did the dredging up of monsters from the deep or, more contemporarily, pictures sent from Mars. Staid scientists would write rapturously of atmospheric dust: what an astonishing thing it is, Wallace mused, that without dust sunsets would be as dull as dishwater, our glorious blue sky would be as black and uniform as ink, and shadows would be so dark and razor-edged as to be as impenetrable as concrete to our sight.

Today the wonders of the atmosphere are often reduced to dry facts that, where they are known at all, are learned by rote by bored schoolchildren. Despite having been forced to swallow them when at school, I still find the workings of the atmosphere fascinating. It connects everything with everything else, and thus performs many services that we take for granted.

It is in our lungs that we connect to our Earth's great aerial bloodstream, and in this way the atmosphere inspires us from our first breath to our last. The time-honoured customs of slapping newborns on the bottom to elicit a drawing of breath, and the holding up of a mirror to the lips of the dying are bookmarks of our existence. And it is the atmosphere's oxygen that sparks our inner fire, permitting us to move, eat and reproduce—indeed to live. Clean, fresh air gulped straight from the great aerial ocean is not just an old-fashioned tonic for human health, it is life itself, and 13.5 kilograms of it are required by every adult, every day of their lives.

The great aerial ocean, indivisible and omnipresent, has so regulated our planet's temperature that for nearly 4 billion years Earth has remained the sole known cradle of life amid an infinity of dead gases, rock and dust. Such a feat is as improbable as the development of life itself; but the two cannot be separated, for the great aerial ocean is the cumulative effusion of everything that has ever breathed, grown and decayed. Perhaps it is the means by which life perpetuates the conditions necessary for its existence. If this is so, two profound questions naturally arise: how can the individual components that comprise life co-ordinate their efforts; and (more immediately relevant to ourselves) what can be said of species that threaten that equilibrium?

In 1979 the mathematician James Lovelock published a book, *Gaia*, that delved deeply into these questions.[116] Lovelock argued that Earth was a single, planet-sized organism, which he named Gaia after an Ancient Greek earth goddess. Anyone who has lived close to nature will recognise the thing Lovelock was describing, but because his arguments seemed mystical they discomfited many scientists.

The atmosphere, Lovelock concluded, is Gaia's great organ of inter-connection and temperature regulation. He describes it as 'not merely a biological product, but more probably a biological construction: not living, but like a cat's fur, a bird's feathers, or the paper nest of a wasp, an extension of a living system designed to maintain a chosen environ-ment'.[116] This notion was considered heretical by many, and until Carl Sagan accepted Lovelock's manuscript for the journal *Icarus*, it faced the prospect of remaining unpublished. In truth, Lovelock had few examples to explain how life might act to regulate Earth's temperature. About the best he could offer was the instance of some micro-organisms that inhabit salt marshes where the salt crystals, by reflecting light back to space, keep them cool. These micro-organisms turn black as winter approaches, thereby absorbing heat and warming Earth.

More important to his argument than such flimsy evidence was a profound paradox. The Sun, like all stars, has become more intense as it has aged. Since life evolved its rays have increased in intensity by 30 per cent, yet the temperature of the surface of our planet has remained relatively constant. A drop of one tenth of 1 per cent in the solar radiation reaching Earth can trigger an ice age; so Earth's long-term climatic stability, Lovelock argued, could not have resulted from mere chance.

One reason biologists were so resistant to the concept of Gaia was that they could not imagine species co-operating globally to achieve such an outcome. Indeed, driven by Richard Dawkins's selfish gene theory, most biology was going in the opposite direction—towards a concept of the world wherein even individual genes were at war with each other. The most devastating rebuttal of the Gaia hypothesis is that it is teleological. Lovelock had asserted that the likelihood of Earth's surface temperature resulting from chance was about the same as surviving a drive through peak-hour traffic blindfolded, to which the biologist W. Ford Doolittle replied:

> I think he is right; the prolonged survival of life is an event of extraordinary low probability. It is however an event which is a prerequisite for the existence of Jim Lovelock and thus for the formation of the Gaia hypothesis...Surely if a large enough number of blindfold drivers launched themselves into rush-hour traffic, one would survive, and surely he, unaware of the existence of his less fortunate colleagues, would suggest that something other than good luck was the cause.[216]

It's a fair enough view, but before accepting it let's look at what evidence in favour of Lovelock has been produced since 1979.

The most compelling proof has to do with the idea that, as life has diversified, Gaia has become better at regulating Earth's temperature. For nearly half of its existence—from 4 billion to around 2.2 billion years ago—Earth's atmosphere would have been deadly to creatures such as us. Back then all life was microscopic—algae and bacteria—and its hold on our planet was tenuous. By around 600 million years ago oxygen levels had increased enough to permit the survival of larger creatures—ones whose fossils can be seen with the unaided eye. These early organisms lived during a period of momentous climate change, when four major ice ages gripped the planet, indicating that back then Earth's thermoregulation was not as effective as it is today. Carbonate deposited in rocks (thus taking CO_2 out of the atmosphere) indicates that there was something odd about the carbon cycle back then. Organic matter was being buried at an unprecedented rate. Maybe the break-up of the early continents opened troughs in the ocean floor that rapidly filled with organic-rich sediment, and this led to a runaway refrigeration of the planet. Whatever the case, with less CO_2 in the atmosphere, Earth began to get very cold. Twice—around 710 million and again at 600 million years ago—Earth crossed a threshold that all but exterminated life, freezing our planet right to the equator.[61]

Whatever its ultimate cause, Earth's deep freeze must have been aided by a powerful mechanism known as Earth's albedo. *Albedo* is Latin for 'whiteness', and of course a snow-covered Earth is a lot whiter than one that isn't. The importance of this can be seen from the fact that one-third of all energy reaching Earth from the sun is reflected back to space by white surfaces. Fresh snow reflects the most light (80–90 per cent), but all forms of ice and snow reflect far more sunlight than does water (5–10 per cent). Once a certain proportion of the planet's surface is bright ice and snow, enough sunlight is lost that a runaway cooling effect is created which freezes the entire planet. That threshold is crossed when ice sheets reach around 30 degrees of latitude.

Around 540 million years ago living things began to build skeletons of carbonate, and to do this they absorbed CO_2 from sea water. This affected CO_2 levels in the atmosphere, and ever since then ice ages have been rare. Only twice—between 355 and 280 million years ago, and for the past 33 million years—have they prevailed. An ingenious theory to explain why this might be so has been put forward by Andy Ridgwell of the University of Riverside, California, and his colleagues.[100] They argue that the evolution of tiny, shell-forming plankton more than 300 million years ago was a crucial step in stabilising Gaia's thermostat. Before that, if Earth's temperature dropped for any reason, ice would form and the level of the ocean would fall, exposing the continental shelves. This in turn disrupted the carbon cycle, allowing the oceans to draw ever greater amounts of CO_2 from the atmosphere, which drove temperatures ever lower. The planktonic calcifiers changed all that because they were not tied to the continental shelves. Instead they floated in the open ocean, so the cycling of carbon through their bodies and into the ocean sediments was not as influenced by exposure of the continental shelves. As a result the oceans were prevented from absorbing too much carbon dioxide from the atmosphere, thereby breaking the self-reinforcing cycle that hitherto had turned a slight chill into a full-blown ice age.

If there was ever a single great advance in the establishment of Gaia, the evolution of the planktonic calcifiers was certainly it; but at around the time they were proliferating other changes were occurring that would also have a profound impact on Earth's thermostat. This was during the Carboniferous Period, when forests first covered the land, and when most of the coal deposits that now feed our industry were laid down. All of the carbon in that coal was once tied up in CO_2 floating in the atmosphere, so those primitive forests must have had an enormous influence on the carbon cycle.

Other evolutionary events are likely to have influenced the carbon cycle but, because most have not been studied in detail, we cannot be sure about whether they refined Gaia's thermostatic control or not. The evolution and spread of modern coral reefs around 55 million years ago drew unimaginable volumes of CO_2 from the atmosphere, further altering Gaia; the evolution and spread of grasses around 6–8 million years ago may have changed things in a very different way. Computer simulations reveal that forests would be far more widespread were it not for grasses and the fire they engender. Forests contain much more carbon than does grass, and they also absorb more sunlight (having a different albedo), and produce more water vapour, which affects cloud formation. All of these things influence Gaia's capacity to regulate temperature.[206] Another likely influence on Gaia's thermostat is the elephant, a great destroyer of forests. Like humans, its original homeland was Africa, and as it spread across the planet around 20 million years ago (only Australia escaped colonisation) it too must have affected the carbon cycle.

Despite the growing sophistication of our understanding of how life works to affect Earth's temperature and chemistry, there is still much debate about Lovelock's Gaia hypothesis. But does it really matter whether Gaia exists or not? I think that it does, for it influences the very way we see our place in nature. Someone who believes in Gaia sees everything on Earth as being intimately connected to everything else, just as are organs in a body. In such a system, pollutants cannot simply be shunted out of sight and forgotten, and every extinction is seen as an act of self-mutilation. As a result, a Gaian world view predisposes its adherents to sustainable ways of living. In our modern world, however, the reductionist world view is in the ascendant, and its adherents often see human actions in isolation. And it is a reductionist world view that has brought the present state of climate change upon us.

This is not to say that a Gaian philosophy inevitably makes for good environmental practice. I frequently hear people say that all will be OK with climate change because 'Gaia will sort it out'. When Lovelock argued that 'there must be an intricate security system to ensure that exotic outlaw species do not evolve into rampantly criminal syndicates' that disrupt Gaia's thermostat, he seems to agree. Yet notwithstanding the destruction of human civilisation through the agency of climate change, it's difficult to imagine just how Gaia would 'sort it out'. And even if she does manage to rid herself of us, we would take so many other species with us that the repair job to Earth's biodiversity would take tens of millions of years.

The eminent biologist John Maynard Smith said of the debate between Gaian adherents and the reductionists that 'It would be as foolish to argue about which of these views is correct as it would be to argue whether algebra or geometry is the correct way to solve problems in science. It all depends on the problem you are trying to solve.'[216] And this is the view I will take here, for the questions I wish to address are more amenable to a Gaian approach than to a reductionist one. So let's use the term Gaia as shorthand for the complex system that makes life possible, while recognising all the while that it may result from chance.

TWO

THE GREAT AERIAL OCEAN

> The great aerial ocean which surrounds us,
> has the wonderful property of allowing the heat-rays from
> the sun to pass through it without its being warmed by them;
> but when the earth is heated the air gets warmed by contact
> with it, and also to a considerable extent by the heat radiated
> from the warm earth because, although pure, dry air allows
> such dark heat-rays to pass freely, yet the aqueous vapour
> and carbonic acid [CO_2] in the air intercept and absorb them.
> Alfred Russel Wallace, *Man's Place in the Universe*, 1903.

If we are to understand climate change we need to come to grips with three important yet widely misunderstood terms. The terms are greenhouse gases, global warming and climate change. Greenhouse gases are a class of gases which can trap heat near Earth's surface. As they increase in the atmosphere, the extra heat they trap leads to global warming. This warming in turn places pressure on Earth's climate system, and can lead to climate change. Likewise it's important to have weather and climate sorted out. Weather is what we experience each day. Climate is the sum of all weathers over a certain period, for a region or for the planet as a whole. And all, of course, are generated in the atmosphere.

The atmosphere has four distinct layers, which are defined on the basis of their temperature and the direction of their temperature gradient. The lowest part of the atmosphere is known as the troposphere. Its name means the region where air turns over, and it is so called because of the vertical mixing of air that characterises it.

The troposphere extends on average to twelve kilometres above the Earth's surface, and it contains 80 per cent of all the atmosphere's gases. Its bottom third (which contains half of all the gases in the atmosphere) is the only part of the entire atmosphere that is breathable. The key thing about the troposphere is that its temperature gradient is 'upside down'—it is warmest at the bottom, and cools by 6.5°C per vertical kilometre travelled. At first sight this appears contrary to common sense, for you would expect the air nearest the sun (the ultimate source of heat) to be the warmest, but this peculiarity accounts for the well-mixed nature of the troposphere—after all, hot air rises. Another peculiarity is that the troposphere is the only portion of the atmosphere whose northern and southern halves (divided by the equator) hardly mix, a characteristic that spares the inhabitants of the Southern Hemisphere the polluted air that limits horizons and dulls panoramas in the more populated north.

The next layer of the atmosphere, known as the stratosphere, meets the troposphere at the tropopause. In contrast to the troposphere, the stratosphere gets hotter as one rises through it. This is because the upper stratosphere is rich in ozone, and ozone captures the energy of ultraviolet light, re-radiating it as heat. Because it is not disturbed by rising hot air, the stratosphere is distinctly layered, and fierce winds circulate within it.

Some fifty kilometres above the surface of Earth lies the mesosphere. At −90°C it's the coldest portion of the entire atmosphere, and above it lies the atmosphere's final layer, the thermosphere, which is a thin dribble of gas extending far into space. There temperatures can reach

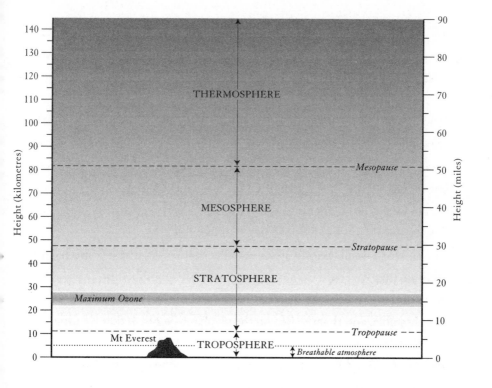

The three major parts of the atmosphere, and their associated boundaries.
Only a small part of the troposphere is breathable air.

1000°C, yet because the gas is so thinly dispersed it would not feel hot to the touch.

The great aerial ocean is composed of nitrogen (78 per cent), oxygen (20.9 per cent) and argon (0.9 per cent). These three gases comprise almost all—over 99.95 per cent—of the air we breathe.[117] And interestingly, its capacity to hold H_2O depends on its temperature: at 25°C water vapour makes up 3 per cent of what we inhale. But as with the watery oceans it's the minor elements—the remaining one-twentieth of 1 per cent—that spice the mix, and some of them are vital to life on this planet. Take, for example, ozone. Its molecules, composed of three oxygen atoms, are scarce even within that tiny minority of spicing gases, which scientists refer to as 'trace' gases. Ozone makes up just ten molecules of every million tossed about in the currents of the great aerial ocean. Yet without the shielding effect of those ten in a million, we would soon go blind, die of cancer or succumb to any number of other problems.

Just as important to our continued existence are the greenhouse gases, of which CO_2 is the most abundant. With fewer than four of every 10,000 atmospheric molecules being CO_2, it can hardly be called common, yet it plays a vital role in keeping us from freezing, and (by its rarity) from becoming overheated. Partly because of it, the average temperature of the surface of our planet is now around 14°C, and ever since complex life first evolved, CO_2 has helped keep it above freezing.

We are so small, and the great aerial ocean so vast, that it seems hardly credible that we could do anything to affect its equilibrium. Indeed, for most of the past century humans have held to the belief that climate is largely stable, and that the flea on the elephant's buttock that is humanity can have no effect. Yet if we were to imagine Earth as an onion, our atmosphere would be no thicker that its outermost parchment skin. Its breathable portion does not even completely cover the surface of the planet—which is why climbers on Mt Everest must wear oxygen masks. And the gases that comprise it are so insubstantial that more gas lies dissolved in the oceans than floats in the atmosphere, and more heat energy is stored near the ocean's surface than in the entire great aerial ocean itself.

To fully understand the atmosphere's vulnerability we must recognise not only its size and gossamer substance, but its dynamism. The air you just exhaled has already spread far and wide. The CO_2 from a breath last week may now be feeding a plant on a distant continent, or plankton in a frozen sea. In a matter of months all of the CO_2 you just exhaled will have dispersed around the planet.[70] Because of its dynamism, the atmosphere is on intimate terms with every aspect of our Earth, from the mantle upward. No volcano belches, no ocean churns—indeed no creature breathes—without the great aerial ocean registering it.

There is one remarkable aspect of the great aerial ocean that has only recently been appreciated—its telekinesis. The last time you heard of telekinesis was probably when Uri Geller was bending spoons, but the term does have a valid scientific definition. It means 'movement at a distance without a material connection', and in the case of the atmosphere telekinesis allows changes to manifest themselves simultaneously in distant regions. Thus, in response to heating or cooling, for example, our atmosphere can at once transform itself from one climatic state into something quite different. This allows storms, droughts, floods or wind patterns to alter on a global level, and to do so more or less at the same time. Telekinetic entities are powerful, but also exquisitely vulnerable to disruption. Our global civilisation is telekinetic, which is why it is such a force in the biosphere, but its telekinesis also explains why regional disruptions—such as wars, famines and diseases—can have dire consequences for humanity as a whole.

The atmosphere is opaque to most forms of radiative energy. Most of us imagine that daylight is the only energy we receive from the Sun, but sunlight—visible light—is only a small band in a broad spectrum of wavelengths that the Sun shoots our way. Light is important to us, of course, for we are creatures of the day whose eyes have evolved to detect wavelengths in just that part of the spectrum. To other wavelengths, the atmosphere is as impenetrable as a brick wall, and it's the gases making up part of that barrier that are the focus of this book: specifically the greenhouse gases, a collection of disparate molecules that share the ability to block long wavelengths of energy. We are more familiar with long wavelengths under the name 'heat energy', and heat is what these gases trap. By doing so, however, they become unstable and eventually release the heat, some of which radiates back to Earth. Greenhouse gases may be rare, but their impact is massive, for by trapping heat near the planet's surface they both warm our world and account for the 'upside down' troposphere.

Some idea of the power greenhouse gases have to influence temperature can be gained by examining other planets. The atmosphere of Venus is 98 per cent CO_2, and its surface temperature is 477°C. Should CO_2 ever reach even 1 per cent of Earth's atmosphere, it would—all other things being equal—bring the surface temperature of the planet to boiling point.[116]

If you want a visceral understanding of how greenhouse gases work, visit New York in August. It's a time of year when the heat and humidity leave those who still trudge the streets in a lather. It feels like such an unhealthy heat—trapped in a crowded, built-up environment of concrete, hard edges, parched bitumen and sticky human bodies—that it is almost insupportable. And the worst of it comes at night, when humidity and a thick layer of cloud lock in the heat. I recall tossing and turning between sweat-soaked sheets in a room near the corner of 9th St and Avenue C. As my eyes became gritty and my skin began to crust up, I could smell the grime of the city's eight million human bodies, along with their refuse and exuviae.

Suddenly I longed to be in a desert—a dry, clear desert where no matter how hot the day, the clear desert skies of night bring blessed relief. The difference between a desert and New York City at night is a single greenhouse gas—the most powerful of them all—water vapour. Reflecting on the fact that water vapour retains two-thirds of all the heat trapped by all the greenhouse gases, I cursed the clouds overhead.[58] But they have a saving grace too. Unlike the other greenhouse gases, water vapour in the form of clouds blocks part of the Sun's radiation by day, keeping temperatures down.

It's testimony to human ignorance that as recently as thirty years ago less than half of the greenhouse gases had been identified and scientists were still divided about whether Earth was warming or cooling. Yet without these molecules our planet would be dead cold—a frigid sphere with an average surface temperature of –20°C. But we *have* known, and for some time, that these gases have been accumulating.

CO_2 is the most abundant of the 'trace' greenhouse gases and it's produced whenever we burn something or when things decompose. In

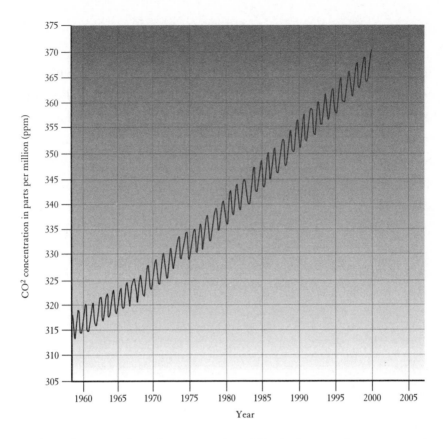

The Keeling curve shows the concentration of CO$_2$ in the atmosphere as measured atop Mt Mauna Loa, Hawaii, between 1958 and 2000. The saw-tooth effect results from the seasonal changes in northern forests, but the inexorable rise is due to the burning of fossil fuels.

the 1950s, a climatologist named Charles Keeling climbed Mt Mauna Loa in Hawaii to record CO$_2$ concentrations in the atmosphere. From this he created a graph, known as the Keeling curve, that is one of the most wonderful things I've ever seen, for in it you can see our planet breathing. Every northern spring as the sprouting greenery extracts CO$_2$ from the great aerial ocean our Earth begins a great inspiration, which is recorded on Keeling's graph as a fall in CO$_2$ concentration. Then, in the northern autumn, as decomposition generates CO$_2$, there is an exhalation which enriches the air with the gas. But Keeling's work revealed another trend. He discovered that each exhalation ended with a little more CO$_2$ in the atmosphere than the one before. This innocent perkiness in the Keeling

curve was the first definitive sign that the great aerial ocean might prove
to be the Achilles heel of our fossil-fuel-addicted civilisation. Looking
back I see that graph as the *Silent Spring* of climate change, for one need
do nothing more than trace its trajectory forward in time to realise that
the twenty-first century would see a doubling of CO_2 in the atmosphere—from the three parts per 10,000 that existed in the early twentieth
century to six. And that has the potential to heat our planet by around
3°C, and perhaps as much as 6°C.

THREE

THE GASEOUS GREENHOUSE

> There prevails an equilibrium in the temperature of the Earth and of its atmosphere…The Earth loses just as much heat by radiation to space and to the atmosphere as it gains by absorption of the sun's rays…I have calculated the mean alteration of temperature that would follow if the quantity of carbonic acid [CO_2] varied from its present mean value.
>
> Svante Arrhenius, *On the Influence of Carbonic Acid in the Air upon the Temperature of the Ground*, 1896.

When scientists first realised that levels of CO_2 in the atmosphere were linked to climate change, some were puzzled. They knew that CO_2 only absorbs radiation at wavelengths longer than about 12 microns (a human hair is around 70 microns thick), and that a small amount of the gas captured all of the radiation available at those bandwidths. Increasing its concentration in experiments seemed to make no real difference to the amount of heat trapped.[70] Besides, there was so little of the gas it seemed inconceivable that CO_2 could change the climate of an entire planet. What scientists did not commonly realise then is that at very low temperatures—such as over the Poles and high in the atmosphere—more heat travels at the bandwidths where CO_2 is most effective. More importantly, they discovered that rather than being the sole agent responsible for climate change, CO_2 acts as a trigger for that potent greenhouse gas, water vapour. It does this by heating the atmosphere just a little, allowing it to take up and retain more moisture, which then warms the atmosphere further. So a positive feedback loop is created, forcing our planet's temperature to ever-higher levels.[70]

Although it is a greenhouse gas, water vapour is also an enigma in the climate change arena, for it forms clouds, and clouds can both reflect light energy and trap heat. By trapping more heat than reflecting light, high thin clouds tend to warm the planet, while low thick clouds have the reverse effect. No single factor contributes more to our uncertainty of future climate change predictions.

Many greenhouse gases are in some way or another generated by human activity. Although scarce, and weak in its capacity to capture heat, CO_2 is very long-lived in the atmosphere: around 56 per cent of all the CO_2 that humans have liberated by burning fossil fuel is still aloft, which is the cause—directly and indirectly—of around 80 per cent of all global warming.[58]

The fact that a known proportion of CO_2 remains in the atmosphere allows us to calculate, in very round numbers, a carbon budget for humanity. Prior to 1800 (the start of the Industrial Revolution) there were about 280 parts per million of CO_2 in the atmosphere, which equates to around 586 gigatonnes (billion tonnes) of CO_2. (To make comparisons easy figures like this relate only to the carbon in the CO_2 molecule. The actual weight of the CO_2 would be 3.7 times greater.) Today the figures are 380 parts per million or around 790 gigatonnes. If we wished to stabilise CO_2 emissions at a level double that which existed before the Industrial Revolution (widely considered the threshold of dangerous change), we would have to limit all future human emissions to around 600 gigatonnes. Just over half of this would stay in the atmosphere, raising CO_2 levels to around 1100 gigatonnes, or 550 parts per million, by 2100. This, incidentally, would be a tough budget for humanity to abide by, for if we use fossil fuels for only another century, that equates to a budget of 6 gigatonnes per year. Compare this with the average of 13.3 gigatonnes of CO_2 that accumulated each year throughout the 1990s (half of this from burning fossil fuel), and the projection that the human population is set to rise mid-century to 9 billion, and you can see the problem.

Even in the long view, this rise is unprecedented. The concentration of CO_2 in the atmosphere in times past can be measured from bubbles of air preserved in ice. By drilling more than three kilometres into the Antarctic ice cap scientists have drawn out an ice-core that spans almost a million years of Earth history. This unique record demonstrates that during cold times CO_2 levels have dropped to around 160 parts per million, and that until recently they never exceeded 280 parts per million. The Industrial Revolution changed that, albeit slowly, for even by 1958 when Keeling began his measurements of CO_2 atop Mauna Loa, it comprised only 315 parts per million.

Australian scientists have recently established that in 2002 and 2003 CO_2 levels rose by 2.54 parts per million per year as opposed to the average increase of 1.8 parts per million per year over the previous decade.[176] It is unclear whether this was simply a 'hiccough' in the trend, or whether rates of accumulation are increasing.

It is our servants—the billions of engines that we have built to run on fossil fuels such as coal, petrol and oil-based fuels, and gas—that play the leading role in manufacturing CO_2. Most dangerous of all are the power plants that use coal to generate electricity. Black coal (anthracite) is composed of at least 92 per cent carbon, while dry brown coal is around 70 per cent carbon and 5 per cent hydrogen.[127] Carbon and oxygen—the components of CO_2—are close neighbours on the periodic table, meaning that they have similar atomic weights. Because two oxygen atoms

combine with one carbon atom to form CO_2, around three and a half
tonnes of the gas is created for every tonne of anthracite consumed. Some
power plants burn through 500 tonnes of coal per hour, and so inefficient
are they that around two-thirds of the energy created is wasted. And to
what purpose do they operate? Simply to boil water, which generates
steam that moves the colossal turbines to create the electricity that powers
our homes and factories. Like the great aerial ocean itself, these
Dickensian machines are invisible to most of us, who have no idea that
nineteenth-century technology makes twenty-first-century gadgets
whirr.

There are around thirty other greenhouse gases in the atmosphere, all
of which are present in trace amounts and whose effect for most purposes is measured by the yardstick of CO_2 (being converted into 'CO_2
units' in scientific equations). Most are so rare that they seem trivial, yet
because they absorb heat at differing wavelengths from CO_2, any increase
in volume is significant. Think of them as glass windows in a ceiling,
each gas representing a different window. As the number of windows
increase, more light energy is admitted into the room, there to be trapped
as heat.[70]

After CO_2, methane is the next most important greenhouse gas.
Although comprising just 1.5 parts per million of the atmosphere, its
concentration has doubled over the last few hundred years. When
measured over a century-long time scale, methane is sixty times more
potent at capturing heat energy than CO_2, but thankfully lasts fewer
years in the atmosphere. Methane is created by microbes that thrive in
oxygenless environments such as stagnant pools and bowels, which is why
it abounds in swamps, farts and belches. It is estimated that methane will
cause 15 to 17 per cent of all global warming experienced this century.
Because it is relatively short-lived, yet is sometimes released suddenly in
vast quantities, methane has played an important role in creating the
positive feedback loops that on occasion have heated our planet.

Nitrous oxide (laughing gas) is 270 times more efficient at trapping heat than CO_2, and although far rarer than methane it lasts 150 years in the atmosphere. Around a third of our global emissions come from burning fossil fuels, and the rest from burning biomass and the use of nitrogen-containing fertilisers. While there are natural sources of nitrous oxide, human emissions now greatly exceed them in volume, and as a result there is 20 per cent more nitrous oxide in the atmosphere than there was at the beginning of the Industrial Revolution.

The rarest of all greenhouse gases are members of the HFC and CFC families of chemicals. These children of human ingenuity did not exist before industrial chemists began to manufacture them. Some, such as the tongue-twisting Dichlorotrifluoroethane, which was once used in refrigeration, are ten thousand times more potent at capturing heat energy than CO_2, and they can last in the atmosphere for many centuries. We shall meet this class of chemicals again later, when we consider the ozone hole.

For the moment, because of its primary importance to climate change, we need to know more about the carbon in CO_2. Both diamonds and soot are pure forms of carbon; the only difference being how the atoms are arranged. Carbon bonds with almost everything non-metallic, which is why life is carbon-based (carbon compounds being diverse enough to enable the complex processes that go on in a body to occur). Carbon is ubiquitous on the surface of planet Earth. It is constantly shifting in and out of our bodies as well as from rocks to sea or soils, and from there to the atmosphere and back again. Its movements are extraordinarily complex and are governed by temperature, the availability of other elements and the activities of species such as ourselves.

Were it not for plants and algae, we would soon suffocate in CO_2 and run out of oxygen. Through photosynthesis (the process whereby plants create sugars using sunlight and water) plants take our waste CO_2 and use it to make their own energy, in the process creating a waste stream of oxygen. It's a neat and self-sustaining cycle that forms the basis of life on Earth. The volume of carbon circulating around our planet is enormous. Around a trillion tonnes of carbon is tied up in living things, while the amount buried underground is far, far greater.[69] And for every molecule of CO_2 in the atmosphere, there are fifty in the oceans.[70]

The places that the carbon goes to when it leaves the atmosphere are known as carbon sinks. You and I and all living things are carbon sinks, as are the oceans and some of the rocks under our feet. Some of these sinks are very large, but they are not infinite, nor is their size steady through time. Over eons, much CO_2 has been stored in the Earth's crust. This occurs as dead plants are buried and carried underground, where they become fossil fuels. It is this buried carbon that allows oxygen to exist in our atmosphere. Should humans somehow be able to take all of that fossil carbon and return it to the atmosphere by burning it, we would use up all the oxygen in our atmosphere.[69] On a shorter time scale, a lot of carbon can be stored in soils, where it forms the black mould beloved of gardeners. Even the uncouth belching of volcanoes (which contains much CO_2) can disturb the climate for long periods of time. And heavenly bodies can also have their impacts, for meteorites and other objects that periodically collide with Earth have so upset the oceans, atmosphere and Earth's crust as to disrupt the carbon cycle.

For the past couple of decades, scientists have been monitoring where the CO_2 that humans produce by burning fossil fuel goes. They can do this because the gas derived from fossil fuels has a unique chemical signature and can be tracked as it circulates around the planet. In very round figures, 2 gigatonnes is absorbed by the oceans and a further 1.5 gigatonnes is absorbed by life on land annually.[73] The contribution made by the land results partly from an accident of history—America's frontier phase of development—that has given some land plants a ravenous hunger for carbon. Mature forests don't take in much CO_2 for they are in balance, releasing CO_2 as old vegetation rots, then absorbing it as the new grows. For these reasons the world's largest forests—the coniferous forests of Siberia and Canada—and the tropical rainforests, are not good carbon sinks, but new, vigorously growing forests are.

During the nineteenth and early twentieth century, America's pioneers cut and burned the great eastern forests, and burned and grazed the western plains and deserts. Then shifts in land use allowed the vegetation to grow back. As a result, most of America's forests are less than sixty years old and are regrowing vigorously, in the process absorbing around half a billion tonnes of CO_2 annually from the atmosphere, and newly planted forests in China and Europe may be absorbing an equal amount. For a few crucial decades these young forests have helped cool our planet by absorbing excess CO_2. But as the Northern Hemisphere's forests and shrublands recover from their mauling at the hands of the pioneers, they will extract less and less CO_2, at the very time that humans are pumping more of it into the atmosphere.

The long-term prospect for forests assisting in the fight against global warming was spelled out in a recent study that examined our planet's carbon budget over two centuries.[74] It demonstrated that there really is only one major carbon sink on our planet, and that is the oceans. They have absorbed 48 per cent of all carbon emitted by humans between 1800 and 1994, while over those same two centuries life on land has actually contributed carbon to the atmosphere.

The world's oceans, however, vary in their ability to absorb carbon. One ocean basin alone, the North Atlantic—which comprises only 15 per cent of the ocean surface—contains almost a quarter of all the carbon emitted by humans since 1800.[74] Even more curiously, it appears that the CO_2 was not being absorbed by the North Atlantic basin, but was dumped there after being absorbed elsewhere. That 'elsewhere' turned out to be the North Sea, a shallow marine basin confined between Great Britain and northern Europe, and which, due to an odd stratification of its waters, allows CO_2 to accumulate in subsurface layers from where it is transported to the North Atlantic. So potent a carbon 'kidney' is this small sea that it has removed 20 per cent of all carbon dioxide ever emitted by humans.[76]

Having just discovered our planet's 'carbon kidney', scientists are worried that changes in ocean circulation brought about by climate change will degrade its effectiveness. There are many ways that this could happen, one of which is best contemplated while consuming a warm can of cola. That intense fizz on opening the can is followed by a bland flatness—indicating that the liquid has quickly released the carbon dioxide that gives it its fizz. Cold drinks hold their fizz longer, and what is true for your can of soft drink is also true for the oceans. Cold sea water can hold more carbon than warm sea water, so as the ocean warms it becomes less able to absorb the gas.

One other critical aspect of sea water, in regard to its capacity to absorb CO_2, is the amount of carbonate it contains. Carbonate reaches the oceans from rivers that have flowed over limestone or other lime-containing rocks, and it reacts with the CO_2 absorbed into the oceans. At present there is a balance between carbonate concentration and the CO_2 absorbed. As the CO_2 concentration increases in the oceans, however, the carbonate is being used up. As a result the oceans are becoming more acid, and the more acid an ocean is the less CO_2 it can absorb.

In July 2004 two researchers, Peter Raymond of Yale University and Jonathan Cole of the Institute of Ecosystem Studies in Millbrook, published findings that offered seemingly good news on this front.[65] They found that, due to increased land degradation and enhanced rainfall in its catchment, the Mississippi River was transporting increasing amounts of carbonate to the oceans. 'These observations have important implications for the potential management of carbon sequestration in the United States,' the authors proclaimed.[65] While it may have seemed that an ever more degraded terrestrial environment offered salvation from our climate woes, a response published a few months later by Klaus Lackner put things in perspective.[66] The extra carbonate carried by the sickening river, Lackner informs us, would be enough to cover America's CO_2 emissions for just thirty-six hours out of every year. Should the phenomenon be true for every river in the world, it would still account for only ten days' worth of the world's emissions per year.

Carbon dioxide in the oceans is also taken up by living things, some of which die and sink, in the process carrying carbon to the sea floor. While living, some of these creatures are vulnerable to the effects of an acidifying ocean, for they are unable to make the carbonate shells that they depend on. All of this means that before the end of this century the oceans are predicted to be taking in 10 per cent less CO_2 than they do today. Indeed scientists have already discovered that the fraction of human-made CO_2 absorbed by the ocean is decreasing.[74] During the 1980s the ocean was taking up around 1.8 gigatonnes of carbon per year, but by the 1990s that had dropped to below 1.6 gigatonnes.[136]

Having seen something of the workings of the atmosphere, its greenhouse gases and the carbon cycle, we can now consider what it all means, and there is no better way to do that than to turn to the work of those ingenious scientists of yesteryear who, without the benefits of computers, satellites or mass spectrometers, combined simple observation and pure reason to deduce that, through our tampering with the atmosphere, the world had a problem.

FOUR

THE SAGES AND THE ONION SKIN

A simple calculation shows that the temper-
ature in the Arctic regions would rise about 8 to 9°C if the
carbonic acid [CO_2] increased 2.5 or three times its present
value...The world's present production of coal reaches in
round numbers 500 millions of tonnes per annum, or 1 ton
per square kilometre of the Earth's surface.

Arvid Gustav Högbom, 'Om Sannolikheton
FöSekulära Forandringar I Atmosfärens Kolsyrehalt', 1894.

The twentieth century opened upon a greatly altered world. Charles
Darwin was eighteen years in the grave, Gregor Mendel's pioneering
studies into genetic inheritance were about to be rediscovered, and the
horse was nearing the end of its tenure as humanity's principal source of
transport. Yet one relic of a heroic, earlier age remained. In his eighth
decade of life Alfred Russel Wallace was still writing with as much
energy and vision as ever. Indeed, when he passed away on the eve of the

Great War, aged ninety, his obituary notice proclaimed that 'he laid aside his pen only to die'.[77] Of all the productions of his twilight years, none rivalled the monumental work that marked his eightieth year. *Man's Place in the Universe* is a lucid yet idiosyncratic book, which attempts to demonstrate that mankind is the pinnacle, the centre—literally the reason for the existence of everything. Its central thrust—along with an attachment to spiritualism and a determined rejection of the benefits of immunisation—caused Wallace to be seen as a heretic in an increasingly orthodox scientific world. And yet for all its foibles *Man's Place in the Universe* is full of insights that resonate with an environmentally aware twenty-first century.

What makes the book so percipient is its author's integrated, holistic way of thinking. It's a similar approach to that of James Lovelock and his Gaia theory; and as was Lovelock, Wallace was struck with the realisation that even slight variations in existing conditions could make Earth uninhabitable. This observation became the book's refrain—that the foul effusions of the Industrial Revolution threatened humanity. As the octogenarian warmed to his subject, so too did his blood rise. 'Let everything give way to this,' he rails as he rallies all humanity to fight the polluters. 'As in a war of conquest or aggression nothing is allowed to stand in the way of victory.'[5]

Wallace was not the first to condemn air pollution, nor the first to foresee its many dangers. *Fumifugium, or the Inconvenience of Aer and Smoak of London Dissipated, Together with Some Remedies Humbly Proposed* was published by the English writer and panjandrum John Evelyn in 1661.[80] As Evelyn pointed out, even at that early date, so vile were the effusions of coal-burning fires that they could be smelled from miles away. London, he wrote, resembled 'the suburbs of Hell'. A few decades later Timothy Nourse published an essay on London's air, in which he wrote that the vapours were eating the city alive, leaving its oldest buildings 'peel'd and fley'd as I may say to the very bones by this hellish and subterraneous Fume'. John Graunt, a London draper who in 1662 compiled the first methodical analysis of London's mortality records, was also concerned.[81] Graunt had nothing to rely on but the records of 'ancient matrons', who bore the onerous responsibility of inspecting all corpses in the city and reporting on the cause of death. Many of their diagnoses are, from today's perspective, incomprehensible; perhaps they even puzzled Graunt, for among causes given are 'Affrighted, Grief, Itch, Piles, Planet, Rising of the Lights', and 'Mother'. This last cause of death relates to a belief, widely held in the seventeenth century, that the organs of the body were rather like the inhabitants of a village. If any were unhappy they could revolt and wander off in search of a better situation. The womb was thought to be particularly liable to upset: if it got too much sex, or perhaps not enough, it was likely to take umbrage and wander towards the neck, where it could cause shortness of breath or even suffocation. A favoured cure for any woman suffering from 'Mother' was to affix a foul-smelling sponge to her suffocating mouth and a sweet-smelling one to her nether regions to entice the womb back into position. Today both 'Rising of the lights' (the 'lights' being lungs) and 'Mother' would be described as diseases of the lungs. As it was, the matrons' reports were enough for Graunt to assert what had been obvious to the public for centuries: lung disorders were a leading cause of death in the metropolis—far more so than in the countryside—and the cause lay with the city's appalling, coal-fuelled pollution.

Frighteningly, little improvement was made to the quality of London's air until after World War II. Indeed, by the time of the Great Smog of 1952, which killed 12,000, many Londoners had developed a perverse pride in their city's filthy air.[117] But Wallace was different. He was apoplectic at the way it stunted and sickened young, growing bodies. Yet his concerns went far beyond that, for he could understand the invisible effects that burning all that coal was having on the systems that keep Earth habitable.

A century before Wallace first drew breath, the brilliant French mathematician Jean Baptiste Fourier was struggling to establish what determines the average temperature of Earth's surface. He wondered why the planet didn't just keep heating up as it is struck by sunlight, until it became as hot as the sun itself. The answer lay in heat radiation, which carries energy back into space at a rate that in a cosmic sense 'balances Earth's books', the result being the average temperature of our planet. But as he calculated the balance between incoming energy from the Sun, and outgoing radiation, he kept getting nonsense results. His workings showed Earth should be an ice-block, frozen solid at –15°C below zero. Then, in a flash of inspiration, Fourier realised that his calculations on heat energy were correct, but that not all of the heat was escaping to space. Something within the atmosphere, he realised, must be keeping heat in. He envisaged that the atmosphere was acting like the glass in a hothouse, letting sunlight in without interference, but then trapping the heat that the Sun's rays generate on reaching the ground.[70]

Today we can explain Fourier's observation thus: the Sun is a very powerful energy source, and the more powerful the source the shorter the wavelengths of energy generated. Most of the Sun's light energy is very short in wavelength. Visible light ranges from 4000 nanometres (.000004 metres or just four-hundredths of a millimetre) to 7000 nano-metres, and at this wavelength energy passes through the atmosphere without warming it. This, along with another important principle, can be demonstrated by visiting a ski field. There the air remains cold on a sunny day, both because the sun does not warm the atmosphere (and there is little water vapour in the cold air to trap any heat) and because the sun's energy is reflected back to space by the snow. When the Sun's rays fall on a darker body, such as skin or a ski glove, however, the rays are absorbed and heat is generated. As your ski glove warms to a toasti-ness greater than its surrounds, the heat energy, which has much longer wavelengths than sunlight, is radiated back into the sky, where it is captured by the greenhouse gases in the atmosphere. And so light passes harmlessly through an atmosphere charged with greenhouse gases, but heat has trouble getting out.

For almost seventy years little was made of Fourier's observation. Then Svante Arrhenius, a Swedish chemist (who received a Nobel prize in 1903), decided to investigate the matter further. In the mid-1890s, around the time he undertook the work, the Swede was suffering a marriage break-up. Desperate for an escape from what must have been a truly unhappy situation, Arrhenius spent up to fourteen hours a day, for a year, in repetitive, mind-numbing calculations. These he had undertaken at the behest of some friends, including the geologist Gustav Högbom, who were obsessed with one of the great questions of the day: what had caused the ice ages? It was a mystery that had gripped the imagination ever since Louis Agassiz proved that much of Europe and North America had once been covered by glacial ice. Back then, mammoth, giant deer and woolly rhinos trudged a landscape where fields of wheat now grew. The transformation had clearly been a grand one, and whoever could take the victor's wreath by explaining how the change occurred would be assured of enduring scientific fame.

Arrhenius was able to demonstrate that a reduction of CO_2 in the atmosphere could have brought on an ice age, but more importantly for our purposes he speculated on how CO_2 levels might influence the Earth in future. He thought that, at nineteenth-century rates of coal-burning, the amount of atmospheric CO_2 might double in 3000 years, bringing balmy conditions to Sweden. This he approved of, thinking only that the process was a little too slow and that it might be hastened by burning more coal. Despite the attraction they may have held for Scandinavians and others afflicted by hard winters, his ideas were soon forgotten. Yet quietly, and without a systematic plan, industry was performing Arrhenius's bidding and increasing the amount of coal burned.

Regardless of these advances climatologists seemed uninterested in the role greenhouse gases play in determining climate. Then in 1938 a steam engineer named Guy Callendar addressed the Royal Meteorological Society in London on the subject. Callendar had an amateur interest in climate trends, and his thorough compilation of statistics, he believed (correctly as it turned out), showed that the world was warming. Moreover, he announced that he knew the cause—the burning of coal and other fossil fuels in industrial machines.[70] Unfortunately, Callendar's prescient study was dismissed by the academicians as the dabbling of an amateur, and soon after the warming trend went into reverse, bringing a temporary end to this line of inquiry.

Around a quarter of a century before Callendar addressed the Royal Meteorological Society, a remarkable reversal of fortune was to lead another climate pioneer to a grand discovery. Milutin Milankovich had spent most of his career practising as a civil engineer in the Austro-Hungarian Empire. Born in what is now Serbia, in 1909 he abandoned his construction work to take up an academic appointment in Belgrade. Soon, however, the turbulent events of the Balkan wars and then World War I intervened, and Milankovich was interned in Budapest, where he was allowed to work in the library of the Hungarian Academy of Sciences. He had already begun to ponder on that great puzzle of his times—the cause of the ice ages—and his internment provided an opportunity to pursue it with a single-mindedness that civilian life could not offer. By the end of the war he had completed a monograph on some aspects of the problem, thus forming a base to which he added over the following decades. In 1941, with the world embroiled in yet another global conflict, Milutin Milankovich was finally ready to publish his great work, *Canon of Insolation of the Ice-Age Problem*.

Milankovich identified three principal cycles that drive Earth's climatic variability. The longest of the cycles concerns the planet's orbit around the Sun. Surprisingly perhaps, Earth's orbit does not describe a perfect circle but an ellipse whose shape changes on a 100,000-year cycle known as Earth's eccentricity. When Earth's orbit is strongly elliptical, the planet is carried both closer to and further away from the Sun, meaning that the intensity of the Sun's rays reaching the Earth varies considerably through the year. At present the orbit is not very elliptical, and there is only a 6 per cent difference between January and July in the radiation reaching

Earth. At times when Earth's orbit is at its most eccentric, however, that difference is 20 to 30 per cent. This is the only cycle that changes the total amount of the Sun's energy reaching Earth, so its influence is considerable.

The second cycle takes 42,000 years to run its course, and it concerns the tilt of Earth on its axis. This varies from 21.8 to 24.4 degrees, and it determines where the most radiation will fall. At the moment the Earth's axial tilt is in the middle of its range. The third and shortest cycle, which runs its course every 22,000 years, concerns the wobble of Earth on its axis. During the course of this cycle, Earth's axis shifts from pointing to the Pole Star to pointing at Vega. This affects the intensity of the seasons. When Vega marks true north, winters can be bitterly cold and summers scorchingly hot.

Only when continental drift brings large parts of Earth's land surface near the Poles can Milankovich's cycles cause ice ages. Then, when the cycles are right, mild summers and harsh winters allow snow to accumulate on the polar lands, until finally it builds into great ice domes.

Even at their most extreme, Milankovich's cycles bring an annual variation in the total amount of sunlight reaching Earth of less than one tenth of 1 per cent. Yet that seemingly trivial difference can cause Earth's temperature to rise or fall by a whopping 5°C. How such small inputs result in such big changes remains a profound mystery, but it is certain that greenhouse gases play a role. Indeed, computer models cannot simulate the onset of an ice age unless atmospheric CO_2 is reduced in the Southern Hemisphere.[58]

Milankovich's *Canon* solved the riddle of the ice ages, but because he published in Serbian it was decades before the world learned of his brilliance. By the time the work was translated into English in 1969, oceanographers had begun to discern, in sediments drawn from the deep ocean floor, empirical evidence of exactly the kind of impacts he had predicted. Today, Milankovich's masterwork is regarded as one of the greatest breakthroughs ever made in climate studies.

With an understanding of greenhouse gases and Milankovich cycles under their belts, climatologists were on their way to grasping why Earth's climate has varied over time; yet there are still other factors to consider.

One is the intensity of radiation emitted from the Sun. About two-thirds of the Sun's rays reaching our planet are absorbed and put to work here, while the remaining third is reflected back into space. It is those captured rays that power our weather and climate, and most life on Earth. Evidence that the Sun is not an unvarying, fiery globe has been with us for a long time. Over 2000 years ago Greek and Chinese astronomers wrote of seeing dark spots on the Sun whose shape and location changed. In April 1612 Galileo, armed with one of the first telescopes, made detailed observations of these sunspots, demonstrating that they were not satellites passing across the surface of the Sun, but that they originate from the star itself. As it happened, Galileo's death in 1642 coincided with a period of exceptionally low sunspot activity which lasted for several centuries, and which may have led to both cold temperatures in Europe and a lack of interest in the phenomenon.

In the nineteenth century serious study of sunspots resumed, and it was discovered that their activity varied on an eleven-year cycle, as well as on a longer cycle of centuries. Sunspots are slightly cooler than the rest of the Sun's surface, yet when there are lots of them, Earth paradoxically seems to warm up. A scarcity of sunspots is thought to account for around 40 per cent of the temperature decrease experienced during the so-called Maunder Minimum of 1645–1715.[63] During this period Europe's temperature plummeted, causing the Thames and the Dutch Inselmeer regularly to freeze over. The role that sunspots have played in these changes, however, is still challenged by some, for despite the coincidence in timing no testable physical mechanism has yet been identified which would allow sunspots to affect the temperature our planet.[117]

Scientists recently acknowledged that variations in solar radiation and greenhouse gas concentrations affect Earth's climate in fundamentally

different ways. This is because solar radiation warms the upper levels of
the stratosphere through the ultraviolet rays that are absorbed by ozone.[78]
Greenhouse gases, in contrast, warm the troposphere, and they warm it
most at the bottom where their concentration is greatest. At the moment
Earth is experiencing both stratospheric cooling (due to the ozone hole)
and tropospheric warming (due to increased greenhouse gases).

This discovery has led to a re-evaluation of some climate shifts, of
which the so-called Medieval Warm Period is most famous. Ever since
H. H. Lamb first wrote of that warmer England of Chaucer, which could
grow its own grapes and make its own wine, the idea that the medieval
Earth was 1–2°C warmer than today has hardly been questioned.[79] It has
in fact become something of a *cause célèbre* of climate change sceptics, for
they have used it to argue that the medieval warming could have had
nothing to do with the burning of fossil fuels, thereby casting doubt on
the link between greenhouse gases and increased temperature. Leaving
this wonky logic aside, the apparent discrepancy was resolved when it
was realised that stratospheric cooling influences circulation in the troposphere, thereby heating and cooling parts of the Earth in a complex, patchy
manner. A survey of global temperature records (from ice-cores, tree-rings
and lake deposits) shows that, if anything, Earth was then overall slightly
cooler (0.03°C) than in the early and mid twentieth centuries, proving that
the idea of a global Medieval Warm Period is bunk.[78]

Greenhouse gases, orbital variations and sunspots can all be thought
of as 'forcing' changes to the temperature of our planet. As scientists began
to contemplate the influence of these forces, and to look to the geological
record to confirm how they worked in the past, they discovered that the
fossil record is characterised by sudden shifts from one steady, long-lasting
climatic state to another. It was as if our planet had reacted in jolts to the
factors that influence climate, and this series of wild shifts drove entire
habitats from one end of a continent to another, causing many extinctions,
yet keeping conditions within bounds tolerable to life.

FIVE

TIME'S GATEWAYS

> The palaeoclimate record shouts out to us
> that, far from being self-stabilising, the Earth's climatic
> system is an ornery beast which overreacts even to small
> nudges.
>
> Wallace Broecker, *Cooling the Tropics*, 1995.

Geology students vexed at having to memorise the divisions of the geological time scale have long resorted to lewd *aides de memoire*. One beloved of the Scots, in whose lands modern geology began, is 'Can Ollie See Down Mike's Pants' Pockets?/Tom Jones Can./Tom's Queer.' The C in 'Can' is for Cambrian, O in 'Ollie' is for Ordovician, S in 'See' is for Silurian, and so on through to our own time, the Quaternary. Having memorised this extensive list, however, students have only learned the basics, for each major division is itself divided into Periods, which are in turn divided into local units. These finest divisions of time are called local units because they are only recognised in limited areas. In North America, for example, the Periods of the Cainozoic Era are divided into local units known as 'North American land mammal ages'. Although they are the smallest divisions on the time scale, many lasted for several million years.

Had life evolved at a steady pace, encountering no setbacks or periods of exceptional opportunity, we would have had no easy way to divide geological time. The divisions of the geological time scale can easily be told apart, though, because of what geologists call 'faunal turnover'—times when species suddenly appear or disappear. We can think of these episodes as time's gateways—occasions when one Age, and often one climate, gives way to the next.

There are just three agents of change sufficiently powerful to open a gateway in time—the shifting of continents, cosmic collisions, and climate-driving forces such as greenhouse gases. All act in different ways, but they drive evolution using the same mechanisms—death and opportunity.

Time's gateways come in three 'sizes'—small, medium and large. The smallest are those opening on brief and local slices of time, of which the 'North American land mammal ages' is a fine example. A common agent of such gateways is migration resulting from continents coming into contact, either because they bump into each other, or because land-bridges form as seas rise and fall, or Earth heats and cools, allowing animals and plants to migrate. In these instances, time's gateways are marked by the sudden arrival of new species, and often the extinction of local competitors.

The medium-sized divisions of time—those separating geological periods—are global in scale, and usually result from factors, such as greenhouse gases, that operate at a global level. In these cases, what you read in the rocks is almost invariably a sorry tale of extinction followed by the slow evolution of new life forms that adapt to the changed

ERA	PERIOD	EPOCH	SIGNIFICANT EVENT	YEARS AGO
				present day
Cainozoic	Quaternary	Holocene	The Long Summer	
				8000
		Pleistocene	Ice Ages *First modern humans*	
				1.8 million
	Tertiary	Pliocene	*First upright human ancestors*	
				5.3 million
		Miocene	*Decline of widespread rainforests*	
				23.8 million
		Oligocene	*Diverse vertebrate communities*	
				33.7 million
		Eocene	*Final separation of Australia from Antarctica*	
			Clathrate release 55 million years ago	55.5 million
		Palaeocene		
			Cretaceous–Tertiary extinction about 65 million years ago	65 million
Mesozoic	Cretaceous		*First flowering plants*	
				145 million
	Jurassic		*First birds*	
				213 million
	Triassic		*First dinosaurs*	
			Permian–Triassic extinction about 251 million years ago	248 million
Palaeozoic	Permian		*First conifers; early reptiles*	
			Ice Ages about 350 to 250 million years ago	286 million
	Carboniferous		*Early amphibians*	
			Late Devonian extinction about 364 million years ago	360 million
	Devonian		*First insects*	
				410 million
	Silurian		*First fish*	
			Ordovician–Silurian extinction about 439 million years ago	440 million
	Ordovician		*Marine invertebrates*	
				505 million
	Cambrian		Cambrian explosion	
				544 million
Proterozoic			Ice Ages about 800 to 600 million years ago	
				2500 million
Achaean			*First life*	
				3800 million
Hadean			*Earth takes form*	
				4500 million

conditions. Time's greatest divisions, however, are those separating Eras.
These are occasions of massive upheaval, when as much as 95 per cent of
all species vanish. Our planet has experienced such massive extinctions
on only five previous occasions, and their causes are mixed. The last time
Earth was afflicted was 65 million years ago, when every living thing
weighing more than 35 kilograms, and a vast number of smaller species,
was destroyed. This is when the dinosaurs vanished, and the cause is
widely believed to have been an asteroid colliding with Earth. Yet that
asteroid devastated only a part of the planet, primarily North America
and northeast Asia. It was the injection of materials into the atmosphere,
thereby changing the climate, which caused the great global dying. We
can therefore think of such extinction as resulting from very rapid climate
change brought about by atmospheric pollution; and CO_2, it turns out,
played a major role in the event.

We know this from the work of palaeobotanists, who have studied
the stomata (tiny breathing holes) in 65-million-year-old fossil leaves.
Those that lived just after the extinction event have far fewer breathing
holes than those living before. This is because carbon dioxide was more
readily available afterwards, and plants needed fewer stomata to get it.
Incidentally, stomata come at a cost, for it is through these holes that the
plant loses water vapour. A study of the precise number of breathing
holes indicates that atmospheric CO_2 rose by thousands of parts per
million, probably because the asteroid collided with limestone-rich rock,
thereby generating huge volumes of CO_2.[59] This instantaneous injection
of greenhouse gas would have caused an abrupt spike in temperature,
and species that could not cope with the increased heat (including many
reptiles) would have perished.

It would be useful to know if any past changes in Earth's climate share
similarities to those we are experiencing now, but unfortunately the
deeper we plumb the geological record the more Old Father Time befud-
dles us by erasing the details. Palaeontologists interested in past climate
change as a key to our future tend to work on rocks 65 million years old

or less, and nowhere is there a greater jackpot than in the deep oceans. Two recent initiatives—the Deep Sea Drilling Project and the Ocean Drilling Program—have retrieved a wealth of information from the ooze and grit that accumulate at the bottom of the sea. Scientists have discovered that innumerable miniature recorders of temperature, salinity and other environmental conditions lie buried in the vertical kilometre or more of fossil-bearing rock intersected by the drills. If you know how to read them, you can play the climatic history of our planet that they encode much like a pianola roll. And, like a pianola, the most captivating rhythms and melodies emerge when information from the cores is fed into the right machines.

The most important of these recorders are the isotopes of oxygen and carbon. Isotopes are atoms with fewer or more neutrons. Oxygen has two stable isotopes, ^{16}O and ^{18}O. Nearly 99.8 per cent of all oxygen on Earth is ^{16}O. The much rarer ^{18}O has two extra neutrons, making it heavier and less likely to evaporate. When the oceans are warm, lots of ^{16}O evaporates, leaving ocean water relatively rich in ^{18}O. Because marine organisms use CO_2 to build their shells, scientists can analyse certain shell fossils for their $^{16}O–^{18}O$ ratio, and thus determine the past temperature. Things become more difficult to interpret during ice ages, however, for then glaciers trap the evaporated ^{16}O in ice at the Poles, making for an especially skewed ratio. This means that, in order accurately to predict past temperatures, geochemists must know whether their samples date from an ice age or not.

Two isotopes of carbon—^{12}C and ^{13}C—are also traced, and these explain oceanic circulation. Plants find it easier to use the lighter isotopes (^{12}C) when they convert sunlight and CO_2 into food, and so blooms of plankton draw large amounts of ^{12}C into the oceans. If those oceans are stratified (as they are today) with layers of warm water near the top, and icy water deeper down, the water cannot circulate and as the plankton die and sink they carry the ^{12}C with them, making the surface layers relatively rich in ^{13}C. But where the cold ocean water wells up from the depths, it carries the ^{12}C with it. Thus, when the ocean was less stratified than today, there was plenty of ^{12}C in the skeletons of surface-dwelling species. Other indicators of past climate include the presence of tropical species, coral growth rings and so forth; and between these indicators and the isotopic studies an exquisitely detailed record can be constructed.

In 2001 James Zachos, of the University of California at Santa Cruz, and his colleagues attempted a grand and ambitious synthesis.[19] Using all of the applicable techniques they analysed oceanic core samples from around the world in an attempt to tell the story of our climate over the past 65 million years. The study proclaimed Milankovich triumphant, for most of the climatic trends Zachos and his colleagues observed had been driven by his cycles, though the opening and closing of seaways and building of mountains also exerted considerable influence. These cosmological and geological factors, however, could not explain three changes that they called climatic aberrations.

These aberrations took place about 55, 34 and 23 million years ago and mark major geological boundaries—the Palaeocene–Eocene, the Eocene–Oligocene and the Oligocene–Miocene. Because the last two boundaries were times of abrupt cooling (during which glaciers advanced for 400,000 and 200,000 years respectively) and were marked by low and declining levels of greenhouse gas, they are of less relevance to our situation today and will not be discussed further.

The oldest climate aberration, that of 55 million years ago, is more relevant to our contemporary situation, for it marks a time when the Earth's surface abruptly heated by 5 to 10°C. Until November 2003 we had little detailed knowledge of this event, for the critical few metres of sediment that mark it seemed to be missing from the sedimentary record. Then the Ocean Drilling Program recovered three cores from the Shatsky Rise (32°N 158°E), a submarine mountain range over two kilometres deep in the north Pacific. Two hundred metres below the sea floor the drill encountered a 25-centimetre-thick layer of ooze, and its analysis revealed an astonishing tale.[19]

The first thing the researchers noted was that the layer sat atop a section of sea floor that had been eaten away by acid, powerful proof that the oceans had turned acidic. It's a trend that we can observe today and which occurs when CO_2 is being absorbed by sea water in large amounts. Not surprisingly, life in the ocean depths was profoundly affected. Foraminifera are tiny marine creatures that play an important role in the oceanic food chain. Because their shells fossilise well, and are readily identified, they often provide the best evidence to explain how climate change has affected ecosystems. Comparisons of foraminifera from above and below the acid-eaten layer reveal that there were massive extinctions in the ocean depths. It seems likely that the entire ecosystem of the deep ocean suffered a severe shock, from the tiny species at the bottom of the food chain to the bizarre deep-sea fish, sharks and squid that form its apex. The surface levels of the ocean were also affected, as witnessed by the arrival of new types of foraminifera inhabiting coasts and open oceans.

On land there is evidence for abrupt changes in rainfall during this period, and the development of a rainfall pattern similar to that seen today in the Amazon Basin, where transpiration of water vapour from plants is the chief source of rainfall.[113] But the thing that really marks this time and changes life on land forever is a series of migrations in which the fauna and flora of Asia stream into North America and Europe, establish themselves, then drive many of the ancient creatures that had persisted there to extinction.

Fifty-five million years ago North America, Asia and Europe were all connected (or nearly connected) by land-bridges that ran through the Arctic Circle, and the abrupt warming made these migration routes briefly accessible to many warmth-loving species. One of the most remarkable things about these changes is that they happened so rapidly: the onset of the warming appears to have occurred over decades or centuries. So what was the cause? In 2004 it was revealed that a mind-boggling 1500 to 3000 gigatonnes of carbon had been injected into the atmosphere.

From a geological perspective the release happened 'instantaneously', meaning that it was so swift that its duration cannot be measured in the sediments. Perhaps it occurred over decades or years, during which time atmospheric concentrations of CO_2 rose from around 500 parts per million (twice the concentration of the last 10,000 years) to around 2000 parts per million.

Norwegian scientists have recently identified a structure that points to where the gas came from.[194] They noticed that 55-million-year-old sediments in the north and central Atlantic lacked any carbonate at all, indicating that acidification of the ocean here was far more severe than elsewhere and suggesting that the gas may have originated nearby. By examining seismic data from the sea off Norway they detected several crater-like structures, up to 100 kilometres across, which reach from deep within Earth to the sediment surface as it was 55 million years ago. At the base of these structures lay narrow ribbons of volcanic rock that had been squeezed up through the Earth's crust.

Putting the pieces together, the Norwegians believe that the climate change of 55 million years ago was driven by a vast, natural gas-driven equivalent of a barbecue. The fuel for the event lay in one of the greatest accumulations of hydrocarbons—mostly in the form of methane gas—that we know about. While largely consisting of fossil fuel buried in sediments, it may also have contained an icy, methane-rich substance known as clathrates that still abounds in the ocean deep. Fuel, however, is of no use without a source of ignition, and those long ribbons of magma provided the spark. We can imagine Earth's crust creaking as the hot tongues of molten rock made their way towards the fuel. Most probably it did not burn, but heated and expanded, forcing its way quickly towards the surface. When it arrived on the sea floor a massive submarine explosion must have ensued, the likes of which the world had never seen.[186] Most of the methane, however, did not reach the atmosphere. Instead it combined with oxygen in the sea water (it was 'burned') leaving only CO_2 to arrive at the surface. With the deep ocean devoid of oxygen, life must have struggled. Then, as CO_2 turned the depths acidic, a cavalcade of creatures, most of which will never be known to us, were force-marched off to extinction. Indeed there is mounting evidence that many of the deep-sea creatures that are with us today evolved after this time.[193]

Because these research findings are so new, details are far from settled. It may be that the vents in the Norwegian Sea released only part of the gas that cooked our planet, and that a positive feedback saw more gas released from clathrates elsewhere as the oceans warmed, bringing on a thermal disaster. Whatever the case, it took at least 20,000 years for the Earth to re-absorb all of the additional carbon, which was apparently soaked up by a bloom of surface plankton.

Because the extinction event of 55 million years ago was caused by a rapid increase in greenhouse gases, it offers the best parallel with our current situation. Yet there are significant differences which mean that the events we and our children will experience will not be a simple re-run of that bygone era. Most important is the fact that Earth has now been in an 'ice house' phase for millions of years, whereas 55 million years ago it was already very warm, with CO_2 levels around twice the level they are today. There were no ice caps then, and presumably fewer cold-adapted species—certainly nothing like narwhals and polar bears. Nor was this warmer world likely to have possessed the wondrous, stratified slices of life we find today on mountains and in the depths of the sea. Thus our modern Earth stands to lose far more from rapid warming than the world of 55 million years ago. Back then the warming closed a geological Period, while we might, through our activities, bring to an end an entire Era.

SIX

BORN IN THE DEEP-FREEZE

> When an icy mantle gradually crept over much of the northern hemisphere, the greater part of the animal life must have been driven southward, causing a struggle for existence which must have led to the extermination of many forms, and the migration of others into new areas. But these effects must have been greatly multiplied and intensified if, as there is good reason to believe, the glacial epoch itself...consisted of two or more alternations of warm and cold periods.
>
> Alfred Russel Wallace, *Man's Place in the Universe*, 1903.

We human beings are, as our scientific name *Homo sapiens* suggests, the 'thinking creatures', and we are in the grand scheme of things very recent arrivals. The Period that gave birth to our species is called the Pleistocene, a word that means the most recent times. The ice age in which we evolved covers the last 2.4 million years, and because of its youth much of the evidence concerning it is still fresh. The first of our kind—moderns in every physical and mental respect—strode about Earth around 150,000 years ago in Africa, and there archaeologists have found bones, tools and the remains of ancient repasts. They had evolved from small-brained ancestors known as *Homo erectus*, who had been in existence for nearly 2 million years. Perhaps the driving force that changed some of 'them' into 'us' was the opportunity offered by the rich shorelines of the African rift lakes, or perhaps the bounty of the Agulhas Current that runs along the continent's southern shores. In such places new foods and challenges may have favoured specialised tool use and selected for high intelligence. Whatever the case, the environment of these distant ancestors was very different from the one we inhabit today, for their world was dominated by an icehouse climate in which the fate of all living things was determined by Milankovich's cycles. Whenever they conspired to expand the frozen world of the Poles, all over the planet chill winds blew and temperatures plummeted, lakes shrank or filled, bountiful sea-currents flowed or slackened, and vegetation and animals alike undertook continent-long migrations.

The genetic inheritance laid down in this world of ice is still with us. A great reduction in the diversity of our genes, for example, tells of a time around 100,000 years ago when humans were as rare as gorillas are today. We could then so easily have vanished, for 2000 fertile adults were all that stood between us and the eternal oblivion of extinction. But soon thereafter the great celestial cycles altered in ways that favoured our species, and by 60,000 years ago small bands of humans had wandered across the Sinai and out into Europe and Asia. By 46,000 years ago they had reached the island continent of Australia, and by 13,000 years ago, as the ice waned for a final time, they discovered the Americas. Now there were millions of us on the planet, and groups thrived from Tasmania to Alaska. Yet for thousands of years these intelligent people, who were like us in every physical and mental way, remained nothing but hunters and gatherers. In the light of our great accomplishments over the past 10,000 years this long period of stasis is an enigma. In order to understand it, we need to investigate the climate that minted our species.

So let's turn to the ice age and the work of those who have devoted their lives to unlocking its secrets.

As we have seen, Earth's sediments are full of climate-recording devices and, the closer we come to our own time, the more information they provide. At their best they yield an annual record of change which includes information on wind direction and speed, the chemistry of the atmosphere, the extent and type of Earth's vegetative cover, the nature of the seasons and the composition and temperature of the oceans—in short, the state of Earth as it was, for example, 5120 years ago.

One of the best sources of information about climate is, in its simplest form, evident to everyone. Look at a piece of timber and you can see, written in its fine texture and growth rings, a story of the way things were when that tree lived. Widely spaced rings tell of warm and bountiful growing seasons when the sun shone and rain fell at the right time. Compressed rings, recording little growth in the tree, tell of adversity when long, hard winters or drought-blighted summers tested life to the limits.

The oldest living thing on our planet is a bristlecone pine growing more than 3000 metres up in the White Mountains of California. More than 4600 years old, it grows in Methuselah Grove alongside many other superannuated specimens. Its precise location is a closely guarded secret because, vulnerable to disturbance, it's been slowly dying for the past 2000 years. Within its trunk this single tree holds a detailed, year-by-year record of climatic conditions in California. Match the pattern from the heart of the Methuselah tree with the rind of a dead stump nearby, and you may pierce time to a depth of 10,000 years. Tree-ring records of this length have now been obtained from both hemispheres, and there is even hope that the great kauri pines of New Zealand, whose timber can lie sound in swamps for millennia, will provide a record spanning 60,000 years of climatic change.

For all its convenience and depth the climate record of the trees is relatively limited in what it can tell us. If you want a really detailed record you must turn to ice—but only in special places does it yield all its secrets. One such place is the Quelccaya ice cap in the high mountains of Peru. There the ice is laid down in a banded annual pattern, each year's snowfall separated by a band of dark dust which is blown up from the deserts below during the winter dry season. Three metres of snow can fall on Quelccaya in a summer, and the falls of subsequent seasons compress it, first to firn (compacted snow) and then to ice. In the process, bubbles of air are trapped, which act as minute archives documenting the condition of the atmosphere. Australian scientists pioneered the techniques allowing methane, nitrous oxide and CO_2 levels to be obtained from these bubbles, each of which reveals its own story about the past conditions of the biosphere. Even the dust is informative, for it tells of the strength and direction of the winds, and of conditions below the ice cap. And isotopes of oxygen in the ice can provide insights concerning the oceans and the distant polar ice caps.

The ice sheets of Greenland and Antarctica yield Earth's longest cores but because ice flows, the older ice is usually compressed and its annual bands disrupted. When the circumstances are right, however, truly spectacular records can be extracted. In the 1990s, European and American teams of researchers were sent to take ice-cores from the Greenland ice plateau. They couldn't agree on a plan so they put down two corers, just far enough apart to ensure that any changes they detected were real and not a localised anomaly. The European team, drilling to the north, had a stroke of great luck, for their core was sunk atop granitic rocks whose radioactivity generated considerable heat. This melted the lowest layers of ice, which prevented the distortion of the layers above, thereby preserving a detailed climatic record extending back 123,000 years. Using this unique record the team was able to show that spectacular shifts in the North Atlantic climate occurred over just five annual ice-layers, and that 115,000 years ago Greenland experienced a hitherto unknown warm phase that was not mirrored in the Antarctic.[183]

In June 2004, when an ice-core over three kilometres long was drawn from a region of the Antarctic known as Dome C (about 500 kilometres from the Russian Vostok base) even more spectacular results were obtained. The recovery of such a long ice-core must count as one of science's greatest triumphs, for drilling through ice is more hazardous than you might imagine. The drill site was bitterly cold: –50°C at the beginning of the drilling season and –25°C in the middle of the Antarctic summer. The drill itself is just ten centimetres wide, and as it grinds its way downward a slender column of ice is separated and drawn to the surface. The first kilometre was especially difficult, for there the ice is packed with air bubbles, and as the core was drawn up these tended to depressurise, shattering the ice into useless shards. Worse, ice chips can clog the drill head, jamming it fast. In the summer of 1998–99 a drill head trapped over a kilometre below the surface forced the abandonment of the hole, leaving the team with no option but to start all over again. This time, as they drilled the three kilometres to the bottom, they stopped after each metre or two to bring the precious core to the surface.

As the team passed the point reached by earlier drilling, the excitement was palpable.[195] 'You know you were getting stuff that had never been seen before,' a team member said, and each kilometre advanced was celebrated with specially warmed champagne. Then, when they were almost at bedrock another problem emerged. Heat from the rocks below was melting the ice, threatening yet another jamming of the drill bit. The final hundred metres was drilled in late 2004, using as a makeshift bit a plastic bag filled with ethanol (to gently melt its way downwards).

The core from Dome C takes us 740,000 years back in time, and as the final few hundred metres are yet to be dated there's the chance that an even longer record will be obtained.[16] This is an enormous development, for it allows us to glimpse how things stood around 430,000 years ago—the last time that the Milankovich cycles brought Earth into a position similar to that which it occupies today. Back then, the ice revealed, the warm (interglacial) period was exceptionally long, suggesting that our planet may have continued to experience mild conditions for a further 13,000 years.[16]

Warm phases—even far briefer ones than the present—were, however, anomalies during the ice age. More typical are cold periods, including the so-called glacial maxima, when the grip of the ice is at its greatest. The last time this happened was between 35,000 and 20,000 years ago. Back then the sea level was more than 100 metres lower than it is today, altering the very shape of the continents; and North America and Europe's most densely inhabited landscapes lay under kilometres of ice. Even regions south of the ice, such as central France, were treeless subarctic deserts, and their growing season of sixty days was an alternation of freezing northerly winds and a few still periods when a stifling haze of glacial dust filled the air.

It's often said that the priority of an agenda is determined by how big a thing is and how fast it's moving, and by the end of the ice age changes were big and moving very fast indeed. So it's no surprise that climatologists are especially interested in the period from around 20,000 to 10,000 years ago—as the glacial maximum began to wane—for over those ten millennia the overall surface temperature of Earth warmed by 5°C—the fastest rise recorded in recent Earth history.

It is worth comparing the rate and scale of change during this period with what is predicted to happen this century if we do not reduce our emissions of greenhouse gases. If we pursue business as usual, an increase of 3°C (give or take 2°C) over the twenty-first century seems inevitable.[189] While the scale of the change is less than that seen at the end of the last glacial maximum, the fastest warming recorded back then was a mere 1°C per thousand years.[192] Today we face a rate of change thirty times faster—and because living things need time to adjust, speed is every bit as important as scale when it comes to climate change.

Despite the keen focus of scientists on this period, details of how the world shifted from glacial maxima to warm interglacial have been slow coming. In 2000, analysis of a core from Bonaparte Gulf in Australia's tropical northwest revealed that 19,000 years ago, over a period of just 100 to 500 years, sea levels rose abruptly by ten to fifteen metres, indicating that the thaw commenced far earlier than anyone had imagined.[83] Because of difficulties dating the sediments this finding was first viewed with suspicion, but in 2004 a second study in the Irish Sea Basin showed a similar but better-dated rise.[84] The fact that the world did not continue warming in consequence was puzzling, but when the immediate cause of the sea rise was identified the reason became clear. The water, it transpired, had come from the collapse of a Northern Hemisphere ice sheet, which poured somewhere between one quarter and two Sverdrups' worth of fresh water into the north Atlantic. The scale of ocean currents is measured in Sverdrups, named after the Norwegian oceanographer Hans Ulrich Sverdrup. A Sverdrup is a very large flow of water— 1 million cubic metres of water per second—and by disrupting the Gulf Stream this influx had profound consequences.

The Gulf Stream transports vast amounts of heat northward from near the equator—almost a third as much as the Sun brings to Western Europe, and that heat is borne in a stream of warm salty water. As it gives up its heat the water sinks because, being salty, it is heavier than the water around it, and this sinking draws more warm, salty water northwards. If the Gulf Stream's saltiness is diluted with fresh water it does not sink as it cools, and no more warm water is drawn northward in its wake.

The Gulf Stream has stopped flowing in the past. Without the heat it brings the melting glaciers begin to grow again and, as their white surface reflects the Sun's heat back to space, the land cools. Animals and plants migrate or die and temperate regions such as central France are plunged into a Siberian chill. The heat, however, does not vanish. Most of it pools around the equator and in the Southern Hemisphere, where it can cause the melting of glaciers in the south, so that the Sun's rays fall on a dark sea surface instead of on ice, and are absorbed. This heats the world from the bottom up, so to speak, and with the Gulf Stream established once more courtesy of growing northern ice, the world enters another cycle of warming.

Somewhere around two Sverdrups of fresh water is required to significantly slow the Gulf Stream, and the geological record confirms that this happened repeatedly between 20,000 and 8000 years ago. Thus the transition from the ice age to the warmth of today was no gentle segue, but instead the wildest of roller-coaster rides, whose high and low points had the sharpness of saw-teeth.

The most famous and well-studied of such spikes is the younger Dryas, named for an alpine flower whose pollen began showing up in unexpected places as a result of a well-documented chill. The sudden freeze began 12,700 years ago, after warming had caused the collapse of a massive ice-dammed lake of melt-water and the redirection of fresh-water flow across the North American continent—from the Mississippi drainage to the St Lawrence. This big freeze lasted for 1000 years, and much of Europe was plunged into full ice age conditions, leaving many parts of the continent uninhabitable. A further cooling event occurred 8200 years ago, and this one caused temperatures over Greenland to drop around 5°C for 200 years. As with the younger Dryas the breaching of an ice dam seems again to have been responsible, with the flow this time being directed into Hudson Bay.[85]

As the crazy see-sawing caused by alternating melting in the Northern and Southern Hemispheres progressed, it drew Earth jerkily yet inexorably towards its present state. And then this climatic madness gave way to the most serene calm. It was as if, says archaeologist Brian Fagan (Professor Emeritus at the University of California, Santa Barbara), a long summer had arrived whose warmth and stability the world had not seen for half a million years.[218]

As a result, all over the world people who had hitherto been sheltering in huts and living hand to mouth independently began to grow crops, domesticate animals and live in settled towns. It's hard to avoid the feeling that the hostile ice age climate and its savage transition to the interglacial had until then stymied this great flowering of creativity and complexity.[184] Indeed, researchers from the University of California at Davis have recently argued that until some 10,000 years ago extreme cold, low levels of CO_2 and great climatic variability made it impossible to grow crops.[17] Then things changed, and only now have we been able to plumb the causes of our recent good fortune. So let's now turn to that ten-millennia-long summer, and to the revolutionary shift in under-standing its origins that is currently under way.

SEVEN

MAKING THE LONG SUMMER

Where the bee sucks there suck I
In a cowslip's bell I lie;
There I couch when owls do cry.
On the bat's back I do fly
After summer merrily:
Merrily, merrily shall I live now
Under the blossom that hangs on the bough
William Shakespeare, *The Tempest*

The long summer that has been the last 8000 years is without doubt *the* crucial event in human history. Although agriculture commenced earlier (around 10,500 years ago in the Fertile Crescent), it was during this period that we acquired most of our major crops and domestic animals, the first cities came into being, the first irrigation ditches were dug, the first words written down, and the first coins minted. And these changes happened not once, but many times in different parts of the world. Before our long summer was 5000 years old, cities had sprung up in Western Asia, East Asia, Africa and central America, and their similarities are astonishing. Whether they were built by Egyptians, Mayans or Chinese, temples, houses and fortifications are all identifiable as such. It is as if the human mind had sheltered a template for the city all along, and was just waiting until conditions permitted to manifest it. These human settlements were ruled by an elite who relied on artisans. In a few societies writing developed, and in even the earliest of these jottings—clay tablets from ancient Mesopotamia—we recognise life as it is lived in a great metropolis.

Until very recently, it was thought that this long summer resulted from a cosmic fluke: Milankovich's cycles, the Sun and Earth were all 'just right' to create a period of warmth and stability of unprecedented length. Just how extraordinary that fluke is becomes apparent when we compare the previous four warm periods. In every case we see not stability but a long, unsteady cooling until a point is reached where Earth plunges into another frigid spell.

Bill Ruddiman, an environmental scientist at the University of Virginia, found nothing in the natural cycles that could account for the stability of our long summer, and so he began to look for a unique factor—something that was operating only in this last cycle, but in none of the earlier ones. That unique factor, he decided, was ourselves, and in doing so he revolutionised another recent development—the endowing on our post-industrial times of its own geological period.

It was Nobel laureate Paul Crutzen (awarded the prize for research into the ozone hole) and his colleagues who first recognised and named this momentous geological event. They called it the Anthropocene—meaning the age of humanity—and they marked its dawn at AD 1800 when methane and CO_2 brewed up by the gargantuan machines of the Industrial Revolution first began to influence Earth's climate.[15] Ruddiman added an ingenious twist to this argument, for he detected what he believes to be human influences on Earth's climate that occurred long before 1800.

Charting the levels of two critical greenhouse gases—methane and CO_2—in air bubbles trapped in the Greenland and Antarctic ice sheets, Ruddiman discovered an anomaly. The ice reveals that until around 8000 years ago the volume of methane in the atmosphere was mostly controlled by Milankovich's 23,000-year-long orbital insolation cycle. This makes sense, for methane is produced in large volume by swamps, so warm, wet times (when swamps abound) produce more methane than dry, cold times.

At the onset of the last insolation cycle, which commenced 8000 years ago, Milankovich's mechanism lost control of methane emissions. Had the insolation cycle controlled them, methane should have commenced declining around 8000 years ago, and gone into a rapid decline by 5000 years ago. Instead, after taking a shallow dip that bottomed out 5000 years ago, methane concentrations begin a slow but emphatic rise. This, Ruddiman argues, is evidence that humans had wrested control of methane emissions from nature, and so we should mark the Anthropocene's dawn as occurring 8000 years ago rather than 200.

It was the beginnings of agriculture—particularly wet agriculture such as that practised in flooded rice paddies in eastern Asia—that tipped the balance, for such agricultural systems can be prodigious producers of the gas. It's fair to note too that farmers of other crops that require swampy conditions were making their own contributions at around this time. Taro agriculture (which involves the creation and maintenance of water-controlling structures), for example, was well under way in New Guinea by 8000 years ago. Even hunter-gatherers may have had a role. Illustrative of their influence is the construction of weirs that transformed vast areas of southeastern Australia into seasonal swamps. These structures were perhaps the most extensive ever created by non-agricultural people, and were used to regulate swamps for the production of eels. Harvested en masse at great gatherings of the tribes, the eels were then dried and smoked to be traded over large distances.[219]

Ruddiman also found evidence in the ice bubbles that the concentration of CO_2 in the atmosphere was being influenced by humans far earlier than first imagined. The pattern of CO_2 concentrations in the atmosphere through the glacial cycles is well established. Basically, CO_2 levels rise rapidly as the glacial stage ends, then begin a slow decline towards the next cold period. Over the last 8000 years atmospheric CO_2 rose from around 160 parts per million to its pre-industrial high of 280 parts per million. If natural cycles were still in control of Earth's carbon budget, Ruddiman states, CO_2 should have stood at around 240 parts per million by 1800. At first glance his argument looks flimsy. After all, early humans would have needed to emit twice as much carbon as our industrial age did between 1850 and 1990—an output only made possible by an unprecedented population using coal-burning machines. The key, notes Ruddiman, is time. Eight thousand years, in human terms at least, is a long span, and as humans cut and burned forests around the globe their activities acted like a hand casting feathers on a set of scales: eventually enough feathers piled up to tip the balance. And thus, posits Ruddiman, the Anthropocene was created.

So delicate was the climatic stability created by humanity over the past 8000 years, argues Ruddiman, that it was still vulnerable to the great cycles of Milankovich; and archaeologist Brian Fagan argues that these cycles could be amplified into truly monumental impacts on human societies. Consider the slight shift in Earth's orbit between 10,000 and 4000 BC, which brought between 7 and 8 per cent more sunlight to the Northern Hemisphere. This enhanced the rainfall of Mesopotamia by 25 to 30 per cent, markedly altering the ratio of rainfall to evaporation, and increasing the overall moisture available to plants sevenfold. What was once a desert was transformed into a verdant plain that supported dense farming communities. After 3800 BC, however, Earth's orbit reverted to its former pattern and rainfall dropped off, forcing many farmers to abandon their fields and wander in search of food.[218]

Brian Fagan's special interest is climate and past civilisations. He believes that the famine-driven wanderers found refuge in a few strategic locations such as Uruk (now in southern Iraq), where irrigation canals branched off the main rivers. Being first in line for river water, places such as Uruk were buffered from changes in rainfall, and here the starving migrants were put to work by a central authority in construction projects such as the maintenance of irrigation canals. Reduced rainfall, Fagan argues, also forced Uruk's farmers to innovate, and so they used, for the first time, ploughs and animals to till fields in a rotation that involved double cropping. With grain production localised around strategic towns, surrounding settlements began to specialise in producing goods such as pottery, metals or fish, which were sold at Uruk's markets for the ever-scarcer grain. Each of these changes led to the development of a more centralised authority, which in turn founded the world's first bureaucrats, whose job it was to tally and distribute the vital grain.

The sum of all of this change was a shift in human organisation, and by 3100 bc Mesopotamia's southern cities had become the world's first civilisations. Indeed the city, Fagan would argue, is a key human adaptation to drier climatic conditions.

Let's return now to Bill Ruddiman's analysis, because it contains several twists in its tail. He sees a clear correlation with times of low atmospheric CO_2 and several plagues caused by the bacterium *Yersinia pestis*—the 'black plague' of medieval times. These epidemics were global in their reach and killed so many people that forests were able to grow back on deserted farmland. In the process they absorbed CO_2, lowering atmospheric concentrations by 5 to 10 parts per million. Global temperatures then fell and periods of relative cold ensued in places such as Europe.

Ruddiman's thesis implies that, by adding sufficient greenhouse gases to keep the Earth 'just right' to delay another ice age, yet not overheating the planet, the ancients performed an act of chemical wizardry. It is as if, at that stage of our development, we were part of Gaia's balance rather than a destroyer of it. According to Ruddiman, however, it was a damn close thing. If a new ice age were to start, we would likely first see evidence of it around Baffin Island in the eastern Canadian Arctic. Haloes of dead lichens around the island's ice caps tell of stillborn ice ages, because what killed these growths was an accumulation of snow that, even a century ago had conditions been fractionally cooler, would have turned to ice and begun the slide into a frigid world. Had the snow not melted, much of the interior of northeast Canada today would host a cloak of ice that each year would grow further southward.

The new ice-core from Dome C challenges Ruddiman's theory because it reveals that, although our current interglacial is different from the past four (which Ruddiman examined), it's similar in some ways to the fifth before our own, which occurred around 430,000 years ago. Then the confluence of Milankovich's cycles and levels of CO_2 was similar to the present, and the warm spell was exceptionally long—26,000 years, as opposed to 12,000 for the others. Only time will tell whether Ruddiman is correct in placing the beginning of the Anthropocene at 8000 rather than 200 years ago. Nonetheless, his analysis is one of the most provocative and stimulating to be published in recent times.

Regardless of its origin, today there are unmistakable signs that the Anthropocene is turning ugly. So great are the changes scientists are detecting in our atmosphere that time's gates appear once again to be opening. Will the Anthropocene become the shortest geological Period on record?

EIGHT

DIGGING UP THE DEAD

We walk on earth,
we look after,
like rainbow sitting on top.
But something underneath,
under the ground.
We don't know.
You don't know.
What do you want to do?
If you touch,
you might get cyclone, heavy rain or flood.
Not just here,
you might kill someone in another place.
Might be kill him in another country.
You cannot touch him.
                    Big Bill Neidjie, *Gagadju Man*, 2001.

Australia's Aborigines live close to the land, and they have a distinctive way of seeing the world. Instead of viewing things like mining, the weather and biodiversity in isolation, they tend to see a whole picture. Big Bill Neidjie was a truly wise elder who spent his youth living a tribal life in intimacy with the land. When he tells us about the impact of mining in his Kakadu country he doesn't talk of the mines, the tailings and the poisoned earth. In just a handful of words he describes the great cycle that runs from disturbing the eternal living dreaming of the ancestors to the catastrophe awaiting unborn generations.

The challenge he throws down—'What do you want to do?'—is discomforting, because by profaning Earth and touching what lies below we have already provided an answer. My country—Bill's country—is pierced through and through with mines of every type, and more coal is drawn from its innards to be shipped off overseas than from any other place on the planet. Because uranium mines have been developed on parts of Big Bill's land that are rich in myth and tradition, he was probably thinking of uranium rather than coal when he composed his epic poem. Yet in it he has intuited the hidden links between mining, climate change and human wellbeing that scientists have groped towards as they seek to understand the greenhouse effect. Bill's challenge remains there to be answered, because we still have a chance to decide our future. But first we need to learn a little of the history, nature and power of that black stone, coal, and of its slippery ally, oil.

Fossil fuels—oil, coal and gas—are all that remain of organisms that, many millions of years ago, drew carbon from the atmosphere. When we burn wood we release carbon that has been out of atmospheric circulation for a few decades, but when we burn fossil fuels we release carbon that has been out of circulation for eons. Digging up the dead in this way is a particularly bad thing for the living to do.

In 2002, the burning of fossil fuels released a total of 21 billion tonnes of CO_2 into the atmosphere. Of this, coal contributed 41 per cent, oil 39 per cent, and gas 20 per cent.[174] These percentages, however, don't reflect the tonnages burned, for some fuels contain more carbon than others. The energy we liberate when we burn these fuels comes from carbon and hydrogen. Because carbon causes climate change, the more carbon-rich a fuel is, the more danger it presents to humanity's future. Apart from the impurities it contains, several of which (such as sulphur and mercury) are potent pollutants, the best black coal is almost pure carbon. Burn a tonne of it and you create 3.7 tonnes of CO_2. The fuels derived from oil are less carbon rich, containing two hydrogen atoms for every one of carbon in their structure. Because hydrogen is a source of energy which produces more heat when burned than carbon (and in doing so produces only water), burning oil releases less CO_2 per unit used than coal. The fossil fuel with the least carbon content is methane, which has just one carbon atom for every four hydrogen atoms. These fuels thus form a stairway leading away from carbon as the fuel for our economy.

The efficiency with which power is generated by burning a fuel is also an important factor in determining how much CO_2 is produced. Even using the most advanced methods (and most coal-fired power plants come nowhere near this), burning anthracite to generate electricity results in 67 per cent more CO_2 emissions than does methane, while brown coal (which is younger, and has more moisture and impurities) produces 130 per cent more.[118] From a climate change perspective, then, there's a world of difference between using gas or coal to power an economy.

Coal is our planet's most abundant and widely distributed fossil fuel. Those in the industry often refer to it as 'buried sunshine', and in a sense this is an accurate description, for coal is the fossilised remains of plants that grew in swamps millions of years ago. In places like Borneo you can see the initial stages of the coal-forming process taking place. There, huge trees topple over and sink into the quagmire, where a lack of oxygen impedes rotting. More and more dead vegetation builds up until a thick layer of sodden plant matter is in place. Rivers then wash sand and silt into the swamp, which compresses the vegetation, driving out moisture and other impurities. As the swamp is buried deeper and deeper in the earth, heat and time alter the chemistry of the wood, leaves and other organic matter. First, the peat is converted to brown coal and, after many millions of years, the brown coal becomes bituminous coal. If further pressure and heat are applied, and more impurities removed, it can finally become anthracite, and at its most exquisite anthracite—in the form of jet—is a beautiful jewel, as pure a carbon treat as a diamond.

Certain times in Earth's history have been more propitious for forming coal than others. The Eocene Period, around 50 million years ago, was one such time. Then, great swamps lay over parts of Europe and Australia, the buried remains of which form the brown coal deposits found there today. Because it is full of impurities, and is often so wet that some of it must be burned just to dry the fuel that enters the furnaces, brown coal is the most polluting of all fuels. It is also unprofitable to trade: if you want coal that will repay the cost of transport, you must turn to bituminous coal or anthracite. Much of the world's anthracite lived during the Carboniferous Period, 360 to 290 million years ago. Named for the immense coal deposits that were then laid down over much of the world, the Carboniferous world was a very different place from the wetlands of today.

If you'd been able to punt through the ancient swamps of that bygone era, instead of swamp cypress and the like, you would have seen gigantic relatives of clubmosses and lycopods, as well as far stranger plants that are now extinct. The scaly, columnar trunks of the *Lepidodendron* grew in dense forests, each trunk two metres in diameter and soaring forty-five metres into the air. They did not branch until near their tips, where a few short, straggly sproutings bore metre-long grassy leaves. In other places the barrel-like *Sigillaria* grew, a bifurcated plant six metres tall, while gigantic horsetail ferns and seed-ferns made up the remainder of the arborescent flora.[87]

There were no reptiles, mammals or birds in those long ago times. Instead, the stifling, humid growth teemed with insects and their kin. The atmosphere then was rich in oxygen, which allowed creatures with inefficient breathing apparatus to grow to immense size. Millipedes grew to two metres in length, and spiders reached a metre across. Thirty-centi-metre-long cockroaches shared the verdure with dragonflies whose wingspans approached a metre, while in the waters below lurked croco-dile-sized amphibians with huge heads, wide mouths and beady eyes. In pilfering the buried bounty of this alien world we have set ourselves free from the limits of biological production in our present age.

The march towards a fossil-fuel-dependent future started in the England of Edward I, though his subjects only reluctantly turned away from the sweet scent of the burning timber that had fed hearths for centuries. The King himself so detested the smell of coal that in 1306 he banned the burning of it in his kingdom, threatening offenders with 'great fines and ransoms'. There are even records of coal burners being tortured, hanged or decapitated (sources don't agree on the punishment—it's possible all three were applied). But England's forests were becoming exhausted, and as the price of timber rose the English became the first Europeans to burn coal on a large scale. For centuries trade in the foul stuff was a monopoly of the Bishop of Durham and the Prior of Tynemouth, whose workers dug it from seams outcropping along the Tyne.

At that time people had no idea what coal was. Many miners believed that it was a living substance that grew underground, and that nothing hastened its multiplication like a good smearing with dung. Perhaps it was the arrival in London of boats full of faecal carbon that caused King Edward's intense dislike of the stuff, but more probably it was associations between coal and disease—or even the devil himself—that caused the ban, for the English were highly suspicious of the black rock. The stench of brimstone (sulphur) that accompanied its burning was an unpleasant reminder of the torments of the infernal regions which, they knew, lay beneath their feet. And its association with disease was most off-putting. Even the Latin word *Carbunculus* (little coal) can mean a disease; and the most frightful symptom of the plague—black swellings of the lymph nodes known as buboes—looked as if they consisted of coal fragments.

Despite this unpromising start, coal would warm, feed and light England's families for over 600 years, and by 1700 a thousand tonnes per day was being burned in the city of London. Throughout the English countryside too, factories depended on coal as their moving power. So great was the demand that an energy crisis soon loomed. England's mines had been dug so deep that they were filling with water, and unless a way of pumping it out could be found, the nation would have to look elsewhere for fuel.

The man who discovered how this might be done was a small-town ironmonger, Thomas Newcomen. His device burned coal to produce steam, which was then condensed to create a vacuum that moved a piston which pumped the water. The first Newcomen engine was installed in a Staffordshire coal mine in 1712. Fifty years later, hundreds of them were at work in mines across the nation, and England's coal production had grown to 6 million tonnes per year.

The ingenious James Watt improved on Newcomen's design and, with the assistance of his able business partner Matthew Boulton, created a market for a new, improved steam engine. Boulton never doubted the enormous potential of his business. When George III asked him how he made his living, he replied, 'I am engaged, your Majesty, in the production of a commodity which is the desire of kings.' When the King asked what he meant, Boulton simply said, 'Power, your Majesty.'[81]

In 1784 Watt's associate and friend William Murdoch produced the first mobile steam engine, transforming coal into a transport fuel, and from that moment on it was clear that the new century—the nineteenth—was to be the century of coal. No other fuel source could rival it for its multiplicity of applications, from cooking to heating, industrial purposes and transport. In 1882 when Thomas Edison opened up the world's first electric light power station in lower Manhattan, electricity production was added to coal's portfolio, and today power production is the last refuge of this rascally fuel.

Despite the inroads made by oil and gas on coal's empire, more coal is burned today than at any time in the past. Two hundred and forty-nine coal-fired power plants are projected to be built worldwide between 1999 and 2009, almost half of which will be in China. A further 483 will follow in the decade to 2019, and 710 more between 2020 and 2030. About a third of these will be Chinese, and in total they will produce 710 gigawatts (710,000 megawatts) of power.[129] The average life of a coal-fired power station is fifty years, and the CO_2 they produce will continue to warm the planet for centuries after they shut down.

If the nineteenth century was the century of coal, the twentieth has been the century of oil. Indeed the dawn of the hydrocarbon age has been put down as 10 January 1901, when the century was not yet a fortnight old.[86] On that day, on a small hill called Spindletop near Beaumont, Texas, Al Hamill was drilling for oil. He had penetrated more than 1000 feet into the sandstone below, and by 10.30 in the morning, disgusted by his lack of success, was about to up stumps. Then, 'with a deafening blast, and a great howling roar, thick clouds of methane gas jetted from the hole. Then came the liquid, a column of it, six inches wide. It rocketed hundreds of feet into the winter sky before falling back to earth as a dark rain.'[86] Although the first purpose-drilled oil well had been put down forty years earlier in Pennsylvania, the discovery of oil in such deep strata was novel. As drilling became more widespread and ever-deeper, such flows became commonplace, ensuring that oil would quickly drive coal from the fields of transport and home heating. The trouble with oil, however, is that there is far less of it than coal, its distribution is patchier and it's harder to find.

Oil is the product of life in ancient oceans and estuaries. It is composed primarily of the remains of plankton—in particular single-celled plants known as phytoplankton.[57] Most of the world's oil reserves are thought to have originated from deep, still, oxygen-poor ocean basins in areas where upwelling brings cold, nutrient-rich bottom-waters to the sunlit surface.[57] In such conditions, the nutrients supercharge the phytoplankton so that they bloom in enormous quantity, and when they die their remains are carried down to oxygen-free depths, where their organic matter can accumulate without being consumed by bacteria. Earth's oceans are vast—more than double the area of the land—so why is the world not literally swimming in oil? Part of the reason is that the ocean crust is always being recycled, but oil is slippery stuff and, unless something obstructs it, it's likely to ooze out of the rocks and dissipate.

The geological process for making oil is as precise as a recipe for making soufflé. First the sediments containing the phytoplankton must be buried and compressed by other rocks. Then, the absolute right conditions are needed to squeeze the organic matter out of the source rocks and to transfer it, through cracks and crevices, into a suitable storage stratum. This stratum must be porous, but above it must lie a layer of fine-grained, impervious rock, strong enough to withstand the pressures that shot the oil and gas high into the air above Spindletop, and thick enough to forbid escape. In addition, the waxes and fats that are the source of oil need to be 'cooked' at between 100–135°C for millions of years. If the temperature ever exceeds these limits, all that will result is gas, or else the hydrocarbons will be lost entirely. As there is no cook tending the great subterranean ovens wherein oil is forged, the creation of oil reserves is the result of pure chance—the right rocks being cooked in the right way for the correct time, usually in a dome-shaped structure where a 'crust' overlies a porous oil-rich level that prevents the oil's escape.

The house of Saud, the Sultan of Qatar and the other opulent principalities of the Middle East all owe their good fortune to this geological accident, for the conditions in the rocks of their region have been 'just right' to deliver a bonanza of oil. Before it was tapped, just one Saudi Arabian oilfield, the Ghawar, held a seventh of the entire planet's oil reserves. And until 1961 the world's oil companies were finding more and more oil every year, much of it in the Middle East. Since then the rate of discovery has dwindled, yet rates of use have gone up. By 1995 humans were using an average of 24 billion barrels of oil per year, but an average of only 9.6 billion barrels was discovered. It's figures such as these that make many analysts think cheap oil is finished, and with the cost persistently above US $40 per barrel, the market is beginning to agree with them. Some analysts are predicting even higher prices, and perhaps shortages, as soon as 2010, which suggests that something else will be needed to power the economies of the twenty-first century.

That 'something else', many in the industry believe, is natural gas, the principal component (around 90 per cent) of which is methane. Only thirty years ago gas supplied just 20 per cent of the world's fossil fuel, while coal supplied 31 per cent and oil nearly half. By the beginning of the twenty-first century, however, gas had supplanted coal in importance and, if current trends continue, by 2025 it will have overtaken oil as the world's most important fuel source. There are proven reserves of gas sufficient to last fifty years. Only our filthiest fuel, coal, holds greater promise by way of reserves. Thus it's looking likely that this will be the century of gas. For now, though, let's examine fossil fuel use, its future growth, and the burden it is already placing on the planet.

The twentieth century opened on a world that was home to little more than a billion people and closed on a world of 6 billion, and every one of those 6 billion is using on average four times as much energy as their forefathers did 100 years before. This helps account for the fact that the burning of fossil fuels has increased sixteen-fold over that period.[70]

Jeffrey Dukes of the University of Utah has proffered an equation as to how humans are supplying that demand.[57] He starts with the observation that all the carbon and hydrogen in fossil fuels was gathered together through the power of sunlight, captured by long-ago plants. By calculating the efficiency with which plant matter is preserved in sediment, the efficiency with which it is converted into fossil fuels, and the efficiency with which we are able to retrieve that fuel, Dukes has concluded that approximately 100 tonnes of ancient plant life is required to create four litres of petrol.

Given the vast amount of sunlight needed to grow 100 tonnes of plant matter, and the prodigious rate at which we are using petrol, coal and gas, it should come as no surprise that over each year of our industrial age, humans have required several centuries' worth of ancient sunlight to keep the economy going. The figure for 1997—around 422 years of fossil sunlight—was typical. Four hundred and twenty-two years' worth of blazing light from a Carboniferous Sun—and we have burned it in a single year.

Reading Dukes's analysis has changed the way I look at the world. Now, as I tread the sandstone pavements around Sydney, I feel the power of long-spent sunbeams on my bare feet. Looking at the rock through a magnifying lens I can see the grains whose rounded edges caress my toes, and I realise that each one of the countless billions has been shaped by the power of the Sun that, over 300 million years ago, drew water from a primordial ocean which then fell as rain on a distant mountain range. Bit by bit the rock crumbled and was carried into streams, until all that remained was rounded grains of quartz. A million times more energy must have gone into creating sand grains than has ever gone into all human enterprise. From the soles of my feet to the top of my sun-warmed head, I instantly know, in the most visceral manner, what Dukes is saying about fossil sunlight: the past is a truly capacious land, whose stored riches are fabulous when compared with the meagre daily ration of solar radiation we receive.

It makes me realise too that the power and seduction of fossil fuels will be hard to leave behind. If humans were to look to biomass (all living things, but in this case particularly plants) as a replacement, we would need to increase our consumption of all primary production on land by 50 per cent. We're already using 20 per cent more than the planet can sustainably provide, so this is not an option. It's for this reason that Dukes's calculation raises a profound question for our society, the significance of which can only be understood by looking at our overall situation on spaceship Earth.

In 1961 there was still room to manoeuvre. In that seemingly distant age there were just 3 billion people, and they were using only half of the total resources that our global ecosystem could sustainably provide. A short twenty-five years later, in 1986, we had reached a watershed, for that year our population topped 5 billion, and such was our collective thirst for resources that we were using *all* of Earth's sustainable production.

In effect, 1986 marks the year that humans reached Earth's carrying capacity, and ever since we have been running the environmental equivalent of a deficit budget, which is only sustained by plundering our capital base.[60] The plundering takes the form of overexploiting fisheries, overgrazing pasture until it becomes desert, destroying forests, and polluting our oceans and atmosphere, which in turn leads to the large number of environmental issues we face. In the end, though, the environmental budget is the only one that really counts.

Between 1800 and 1980 humans produced 244 petajoules of energy (a joule is a unit of energy used to measure gas consumption; a petajoule is a thousand million joules). Such profligate energy use is truly shocking, but consider that, in the two decades 1980–99, you and I and all other humans who existed then, produced 117 petajoules—almost half of the total emitted over the previous 180 years![74]

By 2001 humanity's deficit had ballooned to 20 per cent, and our population to over 6 billion. By 2050, when the population is expected to level out at around 9 billion, the burden of human existence will be such that we will be using—if they can still be found—nearly two planets' worth of resources. But for all the difficulty we'll experience in finding those resources, it's our waste—particularly the greenhouse gases—that is the limiting factor.

Since the beginning of the Industrial Revolution a global warming of 0.63°C has occurred on our planet, and its principal cause is an increase in atmospheric CO_2 from around three parts per 10,000 to just under four. Most of the increase in the burning of fossil fuels has occurred over the last few decades, and nine out of the ten warmest years ever recorded have occurred since 1990.[67] Just how has that fraction of a degree increase affected life on Earth?

2

ONE IN TEN THOUSAND

NINE

THE UNRAVELLING WORLD

The seasons alter: hoary-headed frosts
Fall in the fresh lap of the crimson rose.
William Shakespeare, *A Midsummer
Night's Dream*

Global warming changes climate in jerks, during which climate patterns jump from one stable state to another. Because of the atmosphere's teleki-netic nature, these changes can manifest themselves instantaneously across the globe. The best analogy is perhaps that of a finger on a light switch. Nothing happens for a while, but if you slowly increase the pressure a certain point is reached, a sudden change occurs, and conditions swiftly alter from one state to another.

Climatologist Julia Cole refers to the leaps made by climate as 'magic gates', and she argues that since temperatures began rising rapidly in the 1970s our planet has seen two such events—in 1976 and 1998. These dates are important, for again and again they mark the onset of remarkable phenomena.

The idea that Earth passed through a climatic magic gate in 1976 originated on the faraway coral atoll of Maiana in the Pacific nation of Kiribati. In fact it originated specifically in a Methuselah of a coral—a 155-year-old *Porites*, one of the oldest corals ever found—that lived and grew there. Maiana lies in an important location, for the central Pacific is where El Niños, which are a major climate force across the globe, are first detected. When researchers drilled a section out of this ancient coral they discovered that it contained a detailed record of climate change extending back to 1840.[22] The magic gate itself was manifested as a sudden and sustained increase in sea surface temperature of 0.6°C, and a decline in the ocean's salinity of 0.8 per cent.

Between 1945 and 1955 the temperature of the surface of the tropical Pacific commonly dipped below 19.2°C, but after the magic gate opened in 1976 it has rarely been below 25°C.[3] 'The western tropical Pacific is the warmest area in the global ocean and is a great regulator of climate,' says Martin Hoerling of the Climate Diagnostics Centre at Boulder, Colorado, for among other things it controls most tropical precipitation and the position of the Jet Stream, whose winds bring snow and rain to North America. In 1977 *National Geographic* ran a feature on the crazy weather of the previous year, which included unprecedented mild conditions in Alaska and blizzards in the lower forty-eight states.[175] The immediate cause was a shift in the Jet Stream, but it wasn't just the United States that was affected: changes occurred as far afield as southern Australia and the Galápagos Islands.

Ever since Charles Darwin used the Galápagos Islands' finches to illustrate his theory of evolution by natural selection the region has been a mecca for biologists, who established research stations to monitor its living creatures. Scientists studying the native finch *Geospiza fortis* watched helplessly as the 1977 drought all but exterminated the species on one of the islands. Of the population of 1300 that existed before the drought, only 180 survived, and these were all individuals with the largest beaks, which enabled them to feed by cracking tough seeds. Of those 180 survivors, 150 were males, so when the rains finally came the male finches found themselves facing tough competition for mates. Again, it was those with the biggest beaks that won out. With this double whammy of natural selection sieving out all except those with the very largest beaks, a measurable shift in the beak size occurred on the island population.[68] Darwin's finches are largely defined on the basis of their beak size, for they divide up the islands' ecological niches according to what they can eat, and with nearly two centuries' worth of beak measurements to look back on biologists felt they were witnessing the evolution of a new species.

The 1998 magic gate is also tied up with the El Niño–La Niña cycle, a two- to eight-year-long cycle that brings extreme climatic events to much of the world. During the La Niña phase, which until recently seemed to be the dominant part of the cycle, winds blow westwards across the Pacific, accumulating the warm surface water off the coast of Australia and the islands lying to its north. With the warm surface waters blown westwards, the cold Humboldt Current is able to surface off the Pacific coast of South America, carrying with it nutrients that feed the most prolific fishery in the world, the anchovetta. The El Niño part of the cycle begins with a weakening of tropical winds, allowing the warm surface water to flow back eastwards, overwhelming the Humboldt and releasing humidity into the atmosphere which brings floods to the normally arid Peruvian deserts. Cooler water now upwells in the far western Pacific, and as it does not evaporate as readily as warm water, drought strikes Australia and southeast Asia. When an El Niño is extreme enough, it can afflict two-thirds of the globe with droughts, floods and other extreme weather.

The 1997–98 El Niño year has been immortalised by the World Wide Fund for Nature (now the WWF) as 'the year the world caught fire'. Drought had a stranglehold on a large part of the planet, and so fires burned on every continent, but it was in the normally wet rainforests of southeast Asia that the conflagrations reached their peak. There over 10 million hectares burned, of which half was ancient rainforest. On the island of Borneo 5 million hectares were lost—an area almost the size of the Netherlands.[3] Many of the burned forests will never recover on a time scale meaningful to human beings, and the impact that this has had on Borneo's unique fauna will, in all probability, never be fully known.

Climatologist Kevin Trenberth and his colleagues believe that the 1997–98 event was an extreme manifestation of a more general impact on the El Niño–La Niña cycle by global warming. Ever since 1976 the cycles have been exceptionally long—one would expect such long cycles only once in several thousand years—and there was an imbalance between the phases, with five El Niños, and only two La Niña.[102] Computer-based modelling supports their research, indicating that as greenhouse gas concentrations increase in the atmosphere a semi-permanent El Niño-like condition will result.

The idea that severe El Niño events can permanently alter global climate was first published in 1996, but was considered highly speculative. The 1998 event changed that, for it released enough heat energy to 'spike' the global temperature by around 0.3°C. The reason this occurred appears to be the pool of warm sea water that builds up over the central western Pacific. This is drawn from warm water across the entire Pacific Ocean: in effect the build-up acts as a concentrator and amplifier of the small rises in global temperatures caused by greenhouse gases, and this in turn feeds back to amplify the intensity of the El Niño cycle.

Some of the changes spawned in 1998 were permanent, for ever since then the waters of the central western Pacific have frequently reached 30°C, while the Jet Stream has shifted towards the North Pole. The new climatic regime also seems prone to generating more extreme El Niños, a topic we will return to later. Now it is time to examine how our changing climate has affected various plant and animal populations.

One of the most powerful tools available to researchers wishing to document the response of nature to climate change is the jottings of birdwatchers, fishermen and other nature watchers. Some of these records are very long—one English family recorded the dates of the first frog and toad croaks they heard on their estate every year between 1736 and 1947—and this type of record is of the utmost importance in revealing how things stood as the curtain separating the Anthropocene from our brave new future began to lift.[233] A huge study drawing on such natural history observations was published in 2003 in the journal *Science*, which reveals the immense scale of the shifts now under way. Researchers Camille Parmesan of the University of Texas and her collaborator Gary Yohe made every effort to exclude dubious data, and the most conservative statistical tests were applied to the mass of figures.[27]

Parmesan and Yohe's database has information on more than 1700 historically recorded species, and draws on a body of wildlife observations stretching back to the days of Gilbert White, whose eighteenth-century book *The Natural History of Selborne* was a pioneer of nature writing. The information includes detailed records of the migration, breeding habits and distribution of birds by amateur birdwatchers, the jottings of botanists about the flowering and shooting of plants, and captains' logs from whaling ships. Many records had been kept by clubs and societies, while others were published in little-read journals such as the *Victorian Naturalist*. Indeed, such was the variety and obscurity of the sources that it had deterred previous attempts at collation and interpretation.

Parmesan and Yohe asked two basic questions. Is an underlying trend evident in all of the regions, habitats and organisms documented? And if so, is that trend in the general direction one would expect, given what we know of climate change?

They found that prior to 1950 there is little evidence of any trend, but since that date, right around the globe, a very strong pattern has emerged.

This manifests itself as a poleward shift in species' distribution of, on average, around six kilometres per decade, a retreat up mountainsides of 6.1 metres per decade, and an advance of spring activity of 2.3 days per decade. These trends accord so strongly with the scale and direction of temperature increases brought about by greenhouse gas emissions that Parmesan and Yohe's findings have been hailed as constituting a globally coherent 'fingerprint of climate change'. While such trends might seem small when compared with the rate of change seen over geological time, they are in fact so rapid and decisive it's as if the researchers had caught CO_2 in the act of driving nature polewards with a lash.

One of the most remarkable changes in distribution concerns the tiny marine organisms called copepods, which have been detected up to 1000 kilometres from their usual habitat. More subtle yet still substantial changes have occurred among thirty-five non-migratory species of Northern Hemisphere butterflies, which have extended their ranges northward, some by as much as 240 kilometres, while at the same time becoming extinct in the south of their habitat as it has been made unsuitable to them.[198] Even tropical species are on the move, with Costa Rica's lowland birds extending 18.9 kilometres northward over a twenty-year period.[144]

With so many species relocating, it's inevitable that human changes to the environment will obstruct migration. A striking example of this is provided by Edith's checkerspot butterfly (*Euphydryas editha*). A distinctive subspecies inhabits northern Mexico and southern California, and increased temperatures in spring have caused the plant that its caterpillars feed on—a type of snapdragon—to wilt early, starving the larvae so they cannot pupate. Suitable habitat once abounded to the north, and the population might have migrated if the sprawl of San Diego didn't now stand in its path. With only 20 per cent of its original range now able to support it, without human help the southern subspecies of Edith's checkerspot will not see out the century.[55] Most of the world's more fertile regions are now occupied by human-modified environments, so many,

many more as yet undocumented instances of species and populations facing extinction may already be occurring.

The early onset of spring activity is a distinctive manifestation of climate change. In the bird world the common murre (*Uria aalge*) has begun to lay its eggs an average of twenty-four days earlier each decade over the period its nesting has been studied. In Europe, numerous plant species have been budding and flowering 1.4 to 3.1 days earlier per decade, while their relatives in North America have been doing so 1.2 to two days earlier. European butterflies are appearing 2.8 to 3.2 days earlier per decade, while migrating birds are arriving in Europe 1.3 to 4.4 days earlier per decade.[144]

One of the most important insights yielded by Parmesan and Yohe's study, however, is that not all species are reacting uniformly to climate change. Different species use different cues to initiate events such as breeding and migration, and the capacities of species to adapt to change vary. Thus, as some species shift rapidly, and others are left behind, Parmesan and Yohe warn us that these trends 'could easily disrupt the connectedness among species and lead to…numerous extirpations and possibly extinctions'.[31] This can occur, for example, when a key food item arrives too late to be of use to a predator, or moves too far north for that predator to use it.

A specific instance of this type of difficulty involves the caterpillars of the winter moth (*Operopthera brumata*). Their sole source of food is young oak leaves, which are soft and nourishing enough for them to munch on for only a few weeks. The problem arises because oaks and moths have different cues to tell them when spring has arrived. It is the warming weather that causes the moth's eggs to hatch, but the oaks count the short cold days of winter as their guide for when to put out their leaves. Spring is warmer than it was twenty-five years ago, but the number of cold days in winter hasn't changed. As a result, the winter moths now hatch up to three weeks before the oaks bear their first leaves. Because the caterpillars can survive only two to three days without food, there are now far fewer of them, and those that do survive generally grow faster because there is less competition for food, meaning the birds have less time to find them.

In this illustration, it seems likely that natural selection will act upon the moth to alter the timing of its hatching, but this will occur only through mass mortality of early hatching caterpillars, and for several decades at least we can expect the species to be rare. Whether the birds, spiders and insects that depend on the moths for food can survive the collapse of their food source, however, is another matter. The researcher who discovered the moth's plight, Marcel Visser of the Netherlands Institute for Ecology, believes that the winter moth is just one example among millions. 'If people look for these effects,' he says, 'they will find them everywhere.'[34] If this is true, then concerns for the species at the top of the food pyramid, such as those that prey on winter moths, must be greater, for they are likely to be losing many potential sources of food. Indeed, our concern for the ecosystem as a whole must be greater, for it implies that all around the world the delicate web of life is being torn apart.

Recent studies have documented similar dislocations in aquatic eco-systems. Over the past few decades, breeding newts have been entering European ponds earlier, while frogs have not. This means that the newt tadpoles are well grown when those of the frogs hatch from their eggs, allowing them to eat large numbers of the frogs' young, which is having an impact on frog numbers. Some reptiles face far more direct threats from global warming, for their sex ratios are determined by the temperature at which the eggs are incubated. Biased sex ratios are already being observed in the painted turtle (*Chrysemys picta*), and it has been predicted that if winter temperatures were to rise even slightly above their present high level, the creatures may find themselves with an all-female population. Crocodiles and alligators may also be at risk, for the eggs of the American alligator (a well-studied species) produce only males when hatched at higher than 32°C, and only females when hatched at less than 31°C.[201]

Even more precarious is the case of the tuatara (*Sphenodon* spp.), a unique reptile that is the last of its lineage, and now restricted to some offshore islands in New Zealand. As it is, the poor creature must have a difficult time with reproduction, for the male is the only reptile to lack a penis (they mate by placing their cloacas together), and from mating to the hatching of the young takes two years. If the eggs remain cool, females will result, but under warmer conditions only males will be born. The tuatara lives at a relatively high latitude in environments likely to be dramatically affected by climate change, so its survival hangs in the balance.

A very different climate change impact was recently detected in Lake Tanganyika, Africa, which is one of the world's oldest and deepest fresh-water bodies. Located just south of the equator, it's home to a host of unique species. Like most lakes, its waters are stratified, with the warmest water on top. This can prevent the mixing of the oxygen-rich upper layers with the nutrient-rich ones below, thereby starving plants in the sunlit layers of nutrients, and those in the deeper layers below of oxygen. In the past, the lake's stratification was seasonally broken down by the southeast monsoons, which stirred its waters and drove the spectacular biodiversity.

Since the mid-1970s, however, global warming has so strengthened stratification in the lake (by warming its surface layers) that the monsoons are no longer strong enough to mix the water. As a result, nutrients no longer surface and oxygen no longer penetrates to any depth. Inevitably, the plankton on which most lake life depends has now declined to less than one-third of its abundance of twenty-five years ago. The spectacular spined snail *Tiphoboia horei*, which is found only in the lake, has lost two-thirds of its habitat; today it's found only at depths of 100 metres or less, whereas twenty-five years ago it ventured three times as deep.[139] These changes, scientists warn, are ongoing and threaten a collapse of the lake's entire ecosystem. While from a biodiversity standpoint Lake Tanganyika is one of the world's most important lakes, it's not unique in its vulnerability to climate change. All over the world the surfaces of lakes are warming, thereby preventing the mixing of their waters and threatening the basis of their productivity.

Even remote, seemingly pristine rainforest is being affected by global warming. In areas of the Amazon far distant from any direct human influence, the proportions of trees that make up the canopy is changing. Spurred on by increased CO_2 levels, fast-growing species are powering ahead, crowding out slower-growing species. As the few rapidly growing species shade out their neighbours, the rainforest's biodiversity is diminished, for the birds and other animals that depend on the slower-growing species as food vanish along with their resources. In other rainforests it's been observed that the plants used by herbivores are growing faster, but their leaves are not as nutritious, because despite the elevated CO_2 the plants are unable to obtain increases in other key nutrients. So great is this rate of declining food value that some leaf-eating mammals, such as Australia's rainforest-dwelling possums, are predicted to decline in abundance as a result of this shift.[188]

Changes to tropical biodiversity resulting from rising temperatures haven't always been so subtle. The 1997–98 El Niño ravaged the nations bordering the southwest Pacific. We have already considered the impact it had on the great forests of Borneo, but less understood is how it affected the forests of New Guinea, the world's second largest island.

The region lying east of Wallace's Line and centred on Australia is known as Meganesia, and it has an ancient and distinctive flora and fauna. The richest habitat in all of Meganesia is the mid-montane oak forests of New Guinea, and they reach their finest development between 1500 and 2000 metres elevation in valleys draining north from the watershed in the centre of the island. There, during the oak-fruiting season, the rich humus of the forest floor is littered with large, shiny brown acorns. If you pick one up, you'll most likely discover that it has been chewed, for these forests are home to more species of possum and giant rat than anywhere on Earth, and many of them love nothing more to snack on than acorns.

When I first saw these wondrous forests—in the Nong River Valley north of Telefomin in 1985—they stretched before me into the blue distance, an unbroken, primeval stand of wilderness. Being the first mammalogist ever to work in that area was a rare privilege, for it quickly revealed itself to be home to many unusual species, some of which were unique to the region and entirely unknown to science. One such creature was a greyish, cat-sized possum with large brown eyes, small paws and a short tail, which the Telefol (who sometimes journeyed into the valley to hunt) knew as *matanim*. They, of course, had known about it for thousands of years, but to scientists like me it was new. It proved to be a primitive species whose origins lay near the base of the New Guinean cuscus family tree; and, from what I could gather from discussions with hunters, it had a singular diet in which fig leaves, fruit and the rotten wood of certain trees loomed large.

The Nong isn't the easiest place in the world to reach, so when the opportunity to return came up in 2001, I jumped at it. You might imagine how excited I was, but even before the helicopter landed my spirits had plummeted. The entire valley, along with the surrounding peaks, had been transformed into a vast grove of vegetable tombstones.

Later, my old Telefol friends told me that during the last half of 1997 little or no rain fell, and the cloudless sky cast bitter frosts that killed the forest trees. By New Year, the remains of the forest had been baked to a crisp and its floor lay covered with the leaves of the dead trees. When it came, the fire raced down through the valley and up onto the adjacent peaks. It burned for months, and even a year later it was likely to flare up from moss and dead plant matter buried deep underground.

This sequence of events had utterly devastated the region, driving the wild animals from their haunts and, as testified by the numbers of marsupial jaws kept as trophies, making the last untouched refuges accessible to hunters. On earlier visits I'd noticed that the trophy jaws were rare, for the difficult terrain and dense cover of the forest limited access. Now strings of hundreds of jaws of the larger and rarer creatures such as tree-kangaroos, possums and giant rats hung from hearths, revealing that even mediocre hunters had been assured of success. Was there, hanging among those prizes, I wondered, the jawbones of the very last *matanim* on Earth? It would take years of research to confirm the presence or absence of such a rare and elusive animal. But from what I saw on my visit in 2001, I think that its survival would have to be counted as a miracle.

In tropical and temperate regions the pace of climate change isn't exceptionally rapid, and so far relatively few species have been adversely affected. At the ends of the Earth, however, climate change is occurring now at *twice* the rate seen anywhere else. If we wish to examine the impacts of rapid change—the sort that will affect the entire planet in future—we must venture into that great realm of eternal ice and snow known as the cryosphere.

TEN

PERIL AT THE POLES

A native woman, alone and melancholy in
a hospital room, told [an] interviewer she would sometimes
raise her hands before her eyes to stare at them: 'right in my
hand, I could see the shorelines, beaches, lakes, mountains,
and hills I had been to. I could see the seals, birds and game'.
Another Eskimo, sensing the breakup of his culture's
relationships with the land...told an interviewer it would be
best all around if the Inuit became 'the minds over the land'.
Their minds, he thought, shaped as they were to the
specific contours of the land, could imagine it well enough
to know what to do.

Barry Lopez, *Arctic Dreams*, 1986.

In the final days of 2004, the cities of the world received some astonishing
news: beginning at its northern tip, Antarctica was turning green.
Antarctic hair-grass (*Deschampsia antarctica*) is one of just two kinds of
higher plants that occur south of the 56th degree of latitude. Hitherto it
barely eked out a living as sparse tussocks crouched behind the north face
of a boulder or some other sheltered spot.[146] Over the southern summer
of 2004, however, great green swards of the stuff began to appear, forming

extensive meadows in what was once the home of the blizzard. It's hard to imagine anything more emblematic of the transformations occurring at the polar ends of our Earth. Yet terrestrial changes pale into insignificance when compared to those occurring at sea, for the sea ice is disappearing.

The subantarctic seas are some of the richest on Earth, and there is a genuine paradox here, for that richness exists despite an almost total absence of the nutrient iron. The presence of sea ice somehow compensates for this, for the semi-frozen edge between salt water and the floating ice promotes remarkable growth of the microscopic plankton that is the base of the food chain. Despite the months of winter darkness the plankton thrives under the ice, allowing the krill that feed on them to complete their seven-year life cycle. And wherever there is krill in abundance there are likely to be penguins, seals and great whales. Indeed, so miraculous is the influence of sea ice on plankton, and therefore on krill and the creatures they feed, that there is almost as much difference between the ice-covered and ice-free portions of the Southern Ocean as there is between the sea and the near-sterile Antarctic continent itself.

Dr Angus Atkinson of the British Antarctic Survey is deeply interested in the relationship between plankton, krill, and the mammals that feed on them. Atkinson and his colleagues examined records of krill catches from the fishing fleets of nine countries working in the southwest Atlantic sector of the Southern Ocean.[114] This is the true home of the krill, for 60 to 70 per cent of their total Southern Hemisphere population resides here. Atkinson and his colleagues divided the records into two time slices: 1926–39 and 1976–2003. When they compared the trends in krill abundance over the two periods they found that before 1939, while there was variation year to year, no overall upward or downward trend in abundance was evident. In other words, the krill population was stable. But a very different pattern was seen in the years following 1976. Ever since that date the krill have been in sharp decline, reducing at the rate of nearly 40 per cent per decade. As Atkinson and his colleagues tell us,

'This is not a localised, short-term effect—it relates to around 50 per cent of the [krill] stock and the data span 1926 to 2003...'

As the krill numbers decreased, those of another major grazing species—the jelly-like salps—have increased. Salps, which had previously been confined to more northerly waters, don't require a great density of plankton to thrive; indeed so modest are their dietary needs that they can survive on the meagre pickings offered by the ice-free parts of the Southern Ocean. As far as a whale is concerned, salps are so devoid of nutrients that an ocean stocked to choking point with them is useless. Indeed, none of the Antarctic's marine mammals or birds finds it worthwhile feeding on them. Atkinson's study sums up: 'These changes among key species have profound implications for the Southern Ocean food web. Penguins, albatrosses, seals and whales...are prone to krill shortages.'[114]

Having identified such a major shift, the researchers were anxious to discover what controlled krill numbers. Year to year the population appeared to fluctuate with the extent of sea ice the previous winter; extensive sea ice means plenty of winter food for the krill. Before satellite coverage in the 1980s it was impossible to obtain accurate, direct records of the extent of winter sea ice around the Antarctic. Now, thanks to an ingenious study of (take a breath) methanesulphonic acid preserved in the Antarctic ice, annual changes in sea ice volume can be estimated.[151] The research reveals that the extent of sea ice was stable from 1840 to 1950, but has decreased sharply since, to such an extent that the northern boundary of the ice has shifted southward from 59.3°S to 60.8°S. This corresponds to a 20 per cent decrease in sea ice extent.[2] The reduction in krill numbers coincided so closely with the reduction of sea ice over time as to leave little doubt that climate change is a profound threat to the world's most productive ocean, and to the largest creatures that exist and which feed there.

To gain a sense of the magnitude and rate of change involved, imagine what it would mean for the beasts of the Serengeti if their grasslands had been reduced by 40 per cent each decade since 1976? Or if your own budget had been similarly slashed? Already there are signs that some Antarctic fauna are feeling the pinch. The emperor penguin population is half what it was thirty years ago, while the number of Adelie penguins has declined by 70 per cent.[232] Such studies suggest that in the near future a point will be reached where, one after another, krill-dependent species will be unable to feed. If so, the southern right whales that have only recently begun to return to Australian and New Zealand shores will no longer come, for they do not feed in such temperate waters, and they need to fatten up on winter krill if they are to travel to their birthing grounds. The humpbacks that traverse the world's oceans likewise will no longer be able to fill their capacious bellies, nor will the innumerable seals and penguins that cavort in southern seas. Instead we'll have an ocean full of jelly-like salps, the ultimate inheritors of a defrosting cryosphere.

The Arctic is a region that is almost a mirror image of the south, for while the Antarctic is a frozen continent surrounded by an immensely rich ocean, the Arctic is a frozen ocean almost entirely surrounded by land. It's also home to 4 million people, which means it is better studied.

Most of the Arctic's inhabitants live on its fringe, and it's there, in places such as southern Alaska, that winters are 2°C to 3°C warmer than they were just thirty years ago. Among the most visible impacts of climate change anywhere on Earth are those wrought by the spruce bark beetle. Over the past fifteen years it has killed some 40 million trees in southern Alaska, more than any other insect in North America's recorded history.[170] Two hard winters are usually enough to control beetle numbers, but a run of mild winters in recent years has seen them rage out of control. The spruce budworm is another threat to the trees, with the female budworms laying 50 per cent more eggs at 25°C than at 15°C.[156]

Anything that lives in the treeless Arctic has got to be tough and versatile, and in his wondrous tribute to the polar regions, *Arctic Dreams*, Barry Lopez singles out the collared lemming (*Dicrostonyx hudsonius*) as the symbol of all it takes to survive there. This unassuming creature, he says:

> became prominent in my mind as a...representative of winter endurance and resilience. When you encounter it on the high summer tundra, harvesting lichen or the roots of cotton grass, it rises on its back feet and strikes a posture of hostile alertness that urges you not to trifle. Its small size is not compromising; it displays a quality of heart, all the most striking in the sparse terrain.[200]

Collared lemmings are true offspring of the extreme north, for they survive even on the hostile northern coast of Greenland, and are superbly adapted to life in the cryosphere. They're the only rodents whose coat turns white in winter, and whose claws in this season grow into capacious, two-pronged shovels used for tunnelling through snow. Their population fluctuates on a cycle of around four years, at the end of which their abundance is such that they may migrate en masse in search of food, thereby giving rise to the idea, erroneously propagated, that they commit suicide by running off cliffs.

Despite the hardiness of its inhabitants, the Arctic ecosystem is especially fragile, and subtle changes such as a season with less snow but more rain can have enormous impacts. In 2004 the Arctic Climate Impact Assessment, a report sponsored by nations with an interest in the region, was published.[156] It documents many changes, as well as projecting more to come, and one of its most striking predictions is that, if global warming trends persist, forests will expand northwards to the edge of the Arctic Sea destroying the tundra. Several hundred million birds migrate to these treeless regions to breed, and as the forest encroaches northwards the great flocks will lose out, to be seen less and less often on their migration south. Indeed, the birds look set to lose more than 50 per cent of their nesting habitat this century alone.[156] For the collared lemming the tundra and life itself are inseparable, and the report states that the species will be extinct before the end of this century. Perhaps all that will be left then will be a folk memory of the small, suicidal rodent. But the real tragedy will be that the lemmings didn't jump. They were pushed.

The Arctic Climate Impact Assessment pays special attention to species important to the Arctic's indigenous people, and none is more vital to their lives than the caribou (or reindeer as the species is known in Eurasia). The Peary caribou is a small, pale subspecies found only in west Greenland and Canada's Arctic islands. Autumn rains now ice over the lichens that are the creature's winter food supply, causing many to starve. The number of Peary caribou dropped from 26,000 in 1961 to 1000 in 1997. In 1991 it was classified as endangered, which meant that it couldn't be hunted, thus becoming irrelevant to the Inuit economy.

The Saami people of Finland have noted a similar icing of the caribou's winter food supply, the details of which were given by Heikki Hirvasvuopio to the compilers of the Arctic Climate Impact Assessment report in 2004:

> During Autumn times the weather fluctuates so much, there is
> rain and mild weather. This ruins the lichen access for the
> reindeer. It is very simple—when the bottom layer freezes,
> reindeer cannot access the lichen. This is extremely different from
> previous years. The reindeer has to claw to force the lichen out
> and the whole plant comes complete with bases. It takes extremely
> long for a lichen to regenerate when you remove the base.[156]

Other factors are also acting to deplete the caribou herds. These include changed patterns of snowfall, which blanket food resources, and flooding rivers that kill thousands of calves as they migrate.[156] In short, as climate change advances, it seems that the Arctic will no longer be a suitable habitat for caribou.

If anything symbolises the Arctic it is surely *nanuk*, the great white bear. He is a wanderer and a hunter, and a fair match for man in the white infinity of his polar world. Every inch of the Arctic lies within his grasp: *Nanuk* has been sighted two kilometres up on the Greenland ice cap, he's been found denning at the bottom of Hudson Bay, just 53°N, and purposefully striding the ice within 100 miles of the true pole itself. 'I used to think that the land would stop them,' remarked Canadian polar bear biologist Ray Schweinsburg, 'but I think they can cross any terrain. The only thing that stops them is a place where there is no food.'[200] And for polar bears, having sufficient food to live means lots of sea ice.

Polar bears, it's true, will deign to catch lemmings or scavenge dead birds if the opportunity presents itself, but it's sea ice and *netsik*—the ringed seal that lives and breeds there—that are at the core of the creature's economy. In 1978, an Inuit hunter and his attendant biologist saw a polar bear make a seal kill in open water, but such an event is as rare as a spring blueberry—the exception that proves the rule.[200]

Netsik is the most abundant mammal of the far north and at least 2.5 million of them swim in its berg-cooled seas. Yet at times climatic conditions are such that they simply cannot breed. In 1974 too little snow fell over the Amundsen Gulf for the seals to construct their snow-covered dens on the sea ice. So they left, some travelling as far as Siberia. And the polar bears? Those that had enough fat followed the seals on their long journeys, but many that had not fed well enough the previous season could not keep up and simply starved.

The plight of the harp seals (*Pagophilus groenlandicus*) living in the Gulf of St Lawrence gives us a clear idea of the shape of things to come. Like the ringed seals, they can raise no pups when there is little or no sea ice present—which happened to them in 1967, 1981, 2000, 2001 and 2002.[156] The run of pupless years that opened this century is worrying. When a run of ice-free years exceeds the reproductive life of a female ringed seal—perhaps a dozen years at most—the Gulf of St Lawrence population, which is genetically separate from the rest of the species, will become extinct. Ringed, ribbon and bearded seals also give birth and nurse on the sea ice. Even the mighty walrus lives under the spell of a frozen sea, for the highly productive ice-edge is its prime habitat.

The great bears are slowly starving as each winter becomes warmer than the one before. A long-term study of 1200 individuals living in the south of their range—around Hudson Bay—reveals that they are already 15 per cent skinnier on average than they were a few decades ago. The feeding season has become just too short for the bears to find enough food, and 15 per cent is a lot of body fat to lose before hibernation. With each year, starving females give birth to fewer cubs. Some decades ago triplets were common; they are now unheard of. And back then around half the cubs were weaned and feeding themselves at eighteen months, while today that number is less than one in twenty.[55] Even those females that successfully give birth face dangers unknown of in times past—increasing winter rain in some areas may collapse birthing dens, killing both the mother and cubs sleeping within. And the early break-up of the ice can separate denning and feeding areas; as young cubs cannot swim the distances required to find food, when this happens they will simply starve to death.

As Schweinsburg says, the only thing that stops *nanuk* is a place where there is no food. And in creating an Arctic with dwindling sea ice, we are creating a monotony of open water and dry land where, for *nanuk* at least, there *is* no food. Without a thick fall of snow there is nowhere to make his winter den; and without ice, snow and *nanuk*, what will it mean to be Inuit—the people who named him, and who understand him like no other? When *nanuk* is fit and well-fed he will strip the blubber from a fat seal, leaving the rest to a retinue of camp followers including the arctic fox, the raven, and the glaucous Thayer's and ivory gulls. At certain times and places many of these creatures depend on *nanuk*, for there is no other giver of bounty in this forbidding land.[200] As the Arctic fills with hungry white bears, what will become of these lesser creatures? Some also depend on sea ice, such as the ivory gull and little auk. Indeed the ivory gull has already declined by 90 per cent in Canada over the past twenty years and, if it continues at that rate, will not see out the century. It looks as if the loss of *nanuk* may mark the beginning of the collapse of the entire Arctic ecosystem.

If nothing is done to limit greenhouse gas emissions, it seems certain that sometime this century a day will dawn when no summer ice will be seen in the Arctic—just a vast, dark, turbulent sea. My guess is that the world will not have to wait even that long to be done with *nanuk*, for before the last ice melts the bears will have lost their constellation of den sites, feeding grounds and migration corridors, without which they cannot breed. Perhaps a cohort of elderly bears will linger on, each year becoming thinner than the last. Or perhaps a dreadful summer will arrive, when the denning seals are nowhere to be found. A few ingenious hunters may eke out a living on a diet of lemming, carrion and sea-caught seals, but they'll be so thin that they will not wake from winter's sleep.

The changes we're witnessing at the Poles are of the runaway type, meaning that unless greenhouse gases can be limited—and quickly— there can be no winners among the fauna and flora unique to the region. Instead, we should expect that the realm of the polar bear, the narwhal and the walrus will simply be replaced by the largest habitat on Earth— the great temperate forests of the Taiga, and the cold, ice-free oceans of the north. In areas where forest does not take over, increasing tempera- tures (and thus increasing evaporation) will give rise to polar deserts, for surprisingly large areas of the Arctic receive very little precipitation.[156]

You might think that the encroaching forests, by taking in CO_2 as they grow, would help abate climate change. Scientists estimate that any such gains will be more than offset by the loss of albedo, for a dark green forest absorbs far more sunlight, and thus captures far more heat, than does snow-covered tundra. The overall impact of foresting the world's northern regions will thus be to heat our planet ever more swiftly and, once this has happened, no matter what humanity does about its emissions, it will be too late for a reversal.[191] Any polar bears or seals surviving in zoos, which have been kept in the hope of one day re-creating their icy realm, will remain captive, for after persisting for millions of years the north polar cryosphere will have vanished forever.

ELEVEN

2050: THE GREAT STUMPY REEF?

> I went upon the reef with a party of gentle-
> men, and the water being very clear round the edges, a new
> creation, as it was to us, but imitative of the old, was there
> presented to our view. We had wheat sheaves, mushrooms,
> stags horns, cabbage leaves, and a variety of other forms,
> glowing under water with vivid tints of every shade betwixt
> green, purple, brown, and white; equalling in beauty and
> excelling in grandeur the most favourite *parterre* of the
> curious florist.
>
> Matthew Flinders, *Voyage to Terra Australis*, 1814.

Of all the ocean's ecosystems, none is more diverse nor—as you might guess from the above remarks—more replete with beauty of colour and form than a coral reef; and none, the climate experts and marine biologists tell us, is more endangered by climate change. I've heard this alarming opinion expressed at conferences, and am always struck by the dumb response of the audience to such shocking news. It's as if they either cannot believe it, or the inevitability of bequeathing a world without such wonders to their children puts the matter beyond contemplation.

Can it be that the world's coral reefs really are on the brink of collapse? It's a question of considerable self-interest to humanity, for coral reefs yield around US $30 billion in income each year, mostly to people who have few other resources. Financial loss, however, may prove to be a small thing compared with the loss of the 'free services' that coral reefs provide. The citizens of five nations live entirely on coral atolls, while fringing reefs are all that stand between the invading sea and tens of millions more. Destroy these fringing reefs, and for many Pacific nations you have done the equivalent of bulldozing Holland's dykes.

One of every four inhabitants of the oceans spends at least part of their life cycle in coral reefs. Such biodiversity is made possible by both the complex architecture of the corals, which provide many hiding places, and the lack of nutrients present in the clear, tropical water.

Interestingly, low levels of nutrients can promote great diversity. Consider that in regions with fertile soils and abundant rainfall, only a few plant species can dominate. They are the 'weedy' ones—those that grow most rapidly given optimum sunlight, water and nutrients—and which can thus out-compete the rest. In contrast, where soils are poor, niche specialists—plants that can thrive only within very narrow limits— proliferate, each of which grows best only where specific nutrients are present in specific amounts, and where rain falls at specific times. The best example of this is seen on the infertile sand plains of South Africa's Cape Province, where 8000 species of shrubby flowering plants co-exist in a mix as diverse as that of most rainforests.

The coral reefs are the marine equivalent of South Africa's sand plain flora. And so we can see that nutrients, and disturbances that break down the structure of coral reefs, are the arch-enemy of their diversity, for then only a few weedy species—mostly marine algae—would proliferate.

When Alfred Russel Wallace sailed into Ambon Harbour in what is now eastern Indonesia in 1857, he saw:

one of the most astonishing and beautiful sights I have ever
beheld. The bottom was absolutely hidden by a continuous series
of corals, sponges, actiniae, and other marine productions, of
magnificent dimensions, varied forms, and brilliant colours. The
depth varied from about twenty to fifty feet, and the bottom was
very uneven, rocks and chasms, and little hills and valleys, offer-
ing a variety of stations for the growth of these animal forests. In
and out among them moved numbers of blue and red and yellow
fishes, spotted and banded and striped in the most striking
manner, while great orange or rosy transparent medusae floated
along near the surface. It was a sight to gaze at for hours, and no
description can do justice to its surpassing beauty and interest.[242]

During the 1990s I often sailed down Ambon Harbour, yet saw no
coral gardens, no medusae, no fishes, nor even the bottom. Instead, the
opaque water stank and was thick with effluent and garbage. As I neared
the town it just got worse, until I was greeted with rafts of faeces, plastic
bags, and the intestines of butchered goats.

Ambon Harbour is just one among countless examples of coral reefs
that have been devastated over the course of the twentieth century. Today,
the prevalent practice of overfishing—including fishing with explosives
and poisons—threatens reef survival, for the stability of coral reefs in the
face of climate change utterly depends on the diversity of fishes and other
creatures they shelter.[142] Disturbing reef biodiversity can also lead to
outbreaks of plague species, such as the crown of thorns starfish. Another
problem is the runoff of nutrients from land-based agriculture and
polluted cities which has helped place most of the world's reefs under
threat. Even protected places such as Australia's Great Barrier Reef are
becoming severely degraded, in this case from a fourfold increase in
nutrient and pollution-rich sediments derived from croplands and which
the intense El Niño events and tropical cyclones characteristic of our new
climate carry far out to sea.[142]

Climate change-induced damage to reefs sometimes comes from

unexpected quarters. The 1997–98 El Niño saw the rainforests of Indonesia burn like never before, and for months the air was thick with a smog cloud rich in iron. Before those fires, the coral reefs of southwestern Sumatra were among the richest in the world, boasting more that 100 species of hard corals, including massive individuals over a century old. Then, late in 1997 a 'red tide' appeared off Sumatra's coast. The colour was the result of a bloom of minute organisms that fed on the iron in the smog.[95] Known as dinoflagellates, the toxins they produced caused so much damage it will take the reefs decades to recover, if indeed they ever do.

The smog cloud generated over southeast Asia during the 2002 El Niño was even larger than that of 1997–98—it was the size of the United States. On such a scale smog can cut sunlight by 10 per cent and heat the lower atmosphere and ocean, all of which causes problems for corals.[124] Dinoflagellate blooms are now devastating coastlines from Indonesia to South Korea and causing hundreds of millions of dollars worth of damage to aquaculture. The prospect of recovery for any east Asian coral reefs looks dimmer than ever.[124]

It's the direct impact of higher temperatures, however, that is proving to be the most threatening aspect of climate change to coral reefs. High temperatures lead to coral bleaching, and to understand that phenomenon we need to examine a reef far from human interference, where warm water alone is causing change. There are, thankfully, some reefs protected by remoteness and size, with no pollution, fishermen or tourists. Myrmidon Reef, lying far off the coast of Queensland, sees almost nothing of humans. Every three years scientists from the Australian Institute of Marine Science in Townsville survey it, and when they did so in 2004 they took along environmental writer James Woodford. He described Myrmidon as looking 'as though it's been bombed'. This was the result of the reef crest being severely bleached, leaving a forest of dead, white coral. Only on the deeper slopes did life survive.[54]

Coral bleaching occurs whenever sea temperatures exceed a certain threshold. Where the hot water pools the coral turns a deathly white. If the heating is transient the coral may slowly recover, but when the heat persists it dies. The phenomenon represents the dissolution of a partnership, for the organisms that make up the world's reefs and atolls are in fact two living things in one. The larger partner in this ecological merger is a pale, sea-anemone-like creature known as a polyp. It gains its greenish, red or purplish hue from a lodger—a type of algae known as zooxanthellae. Under normal circumstances the relationship is a happy, symbiotic one: the coral polyp provides a home and some nourishment to the algae, while the algae provides the polyp with food from photosynthesis. As the temperature of the sea water rises, however, the algae's ability to photosynthesise is impaired, and it costs the polyp more to maintain its partner than it gets in return. As in many a failing relationship, this unequal situation leads to a split, though precisely how the polyp ejects the algae (if it does not leave under its own volition) remains a mystery. If temperatures remain high for a month or two, without their algae the polyps starve to death, leaving a skeletal reef which will eventually become overgrown with soft corals and green algae.

Coral bleaching was little heard of before 1930, and it remained a small-scale phenomenon until the 1970s. It was the 1998 El Niño that triggered the global dying. Some coral reefs were studied intensively both before and after this event, which taught scientists a great deal. In the Indian Ocean, the Scott and Seringapatam reefs were severely affected with bleaching to a depth of thirty metres. Prior to 1998, the percentage of hard coral cover on these reefs was a healthy 41 per cent, then dropped to 15 per cent. On Scott Reef there has been a complete failure of coral recovery since; Seringapatam is recovering slowly.

The Great Barrier Reef is the most vulnerable reef in the world to climate change and, due to higher temperatures near the coast and the debilitating impact of pollution, the corals growing nearer the shoreline were harder hit than those on the outer reef. In all, 42 per cent of the Great Barrier Reef bleached in 1998, with 18 per cent suffering permanent damage. In 2002, with the renewal of El Niño conditions, a pool of warm water around half a million square kilometres developed over the reef. This triggered another massive bleaching event that on some inshore reefs killed 90 per cent of all reef-forming corals, and left 60 per cent of the Great Barrier Reef complex affected. In the few patches of cool water which remained, however, the coral was undamaged.

A survey conducted in 2003 revealed that live coral cover had dropped to less than 10 per cent on half of the reef's area, with large declines evident even in the healthiest sections. Public outrage made political action inevitable, and the Australian government announced that 30 per cent of the reef would be protected. This meant that commercial fishing would be banned, and other human activities severely curtailed, in the newly protected zone. But it is not fishing or tourists that are killing the reef, that is being done by spiralling CO_2 emissions.

Australians emit more CO_2 per capita than any nation on Earth. If the Australian government was truly serious about saving the reef, it would take action in both energy policy and international engagement. Instead, in 2004 the government released its long-awaited energy policy, which enshrined coal at the centre of the nation's energy generation system.

In 2002 a panel of seventeen of the world's leading coral reef researchers warned in an article in *Science* that 'Projected increases in CO_2 and temperature over the next fifty years exceed the conditions under which coral reefs have flourished over the past half-million years'. By 2030, they say, catastrophic damage will have been done to the world's reefs, and by 2050 even the most protected of reefs will be showing massive signs of damage.[97] The message was reinforced in October 2002 when fifteen of the world's greatest authorities on coral reefs met in Townsville, Queensland, to discuss the plight of the Great Barrier Reef.[41] According to reef scientist Dr Terry Done a further rise of 1°C in global temperature would see 82 per cent of the reef bleached; 2°C increase 97 per cent, and 3°C 'total devastation'.[54] Because it takes the oceans around three decades to catch up with the heat accumulated in the atmosphere, it may well be that four-fifths of the Great Barrier Reef is one vast zone

Gobiodon *species C. This small fish is a native of Papua New Guinea.*
The destruction of its reef habitat means it is now restricted to a single coral head.

of the living dead—just waiting for time and warm water to catch up
with it.

Extinctions caused by climate change are almost certainly under way
on the world's reefs, and a tiny species of coral reef-dwelling fish known
as *Gobiodon* species C may be emblematic of them.[96] Most of the habitat
used by this diminutive creature was destroyed by coral bleaching and
associated impacts during the 1997–98 El Niño, and it can now be seen
only on one patch of coral in one lagoon in Papua New Guinea. 'Species
C' indicates that it has not yet been formally named, and such is its
tenuous situation that extinction may occur before its scientific baptism.
We know about *Gobiodon* species C only because a scientist interested in
the genus has spent long months documenting changes in the abundance
of this fish that others might not notice. So great is the diversity of coral
reefs, and so few are the marine biologists that study them, that it isn't
an exaggeration to say that we need to multiply the loss of this one little
fish a thousandfold to gain a sense of the cascade of extinctions that is, in
all likelihood, occurring right now.

Yet despite the enormous damage already evident on the world's coral reefs, some scientists are hopeful that the reefs may yet survive climate change. If we'd been able to visit Australia's Great Barrier Reef 15,000 years ago, they point out, we would have seen little more than a raised line of limestone hills separating a coastal plain from the sea. At that time every coral reef in existence today was high and dry, for the ocean was 100 metres lower than at present. The major hard coral families that make up the reefs were in existence before the end of the Cretaceous Period 65 million years ago, when an asteroid struck the planet and devastated global ecosystems. Just how they survived is unclear, though it was almost certainly only in special refuges. Some scientists think that those survivors altered the chemistry of their skeletons; others argue that, for a time, they did away with skeletons altogether. Corals may be forced to such extremities again in the future, for as CO_2 accumulates in the atmosphere and then diffuses into the ocean, it turns the seas acidic and prevents the coral organism from secreting its hard skeleton.

This history suggests it is possible that some individual coral species could survive, while the overall biodiversity of reefs may not. In order to know which conditions the full diversity of coral reefs can tolerate, we need to consider the earliest evidence we have for the life that swarms on today's reefs. The best place to do this is on a verdant hill called Monte Bolca, near the Italian city of Verona, where finely laminated deposits of limestone packed with the bodies of ancient marine fish have been mined since the sixteenth century.

Fifty million years ago the region around Verona was a lagoon behind a coral reef, and when the reef fish died they were washed into its still waters, where their brilliantly coloured bodies drifted into the oxygenless bottom layers. Without oxygen there can be no decay, and at Monte Bolca the preservation is so exquisite that some of the colour patterns of those long-dead fish can still be seen. Scientists have identified 240 species in the deposits, and among them are the ancestors of many of the fish that inhabit the world's modern coral reefs. The presence of so many fishes so early in the geological record suggests a rapid radiation of species following a catastrophe. Given what we know of the methane-fuelled climatic cataclysm of 55 million years ago, it seems possible that this event devastated earlier coral reef fish—and in all likelihood the reefs themselves—and that in the aftermath the coral reef communities we know today took their place.

There are two ways that the species which constitute coral reefs might survive the looming threat of climate change: by adapting or migrating. Recent research has found that some types of zooxanthellae that live inside the polyps can tolerate higher temperatures than others. One algal form, known as *Symbiodinium* strain D, is particularly good at tolerating warm water, but because it's not as efficient at producing food from sunlight as its low-temperature cousins, it is today relatively rare.[98] On reefs destroyed by bleaching, however, its abundance has increased. If corals can adapt in this way there is hope that some of them, and perhaps reefs, will survive in the locations where they grow right now.[99] Yet the extent of adaptation would need to expand many times over and occur swiftly to save the majority of coral reefs from devastation.

Another escape route may lie in corals migrating south to cooler waters. In the case of the Great Barrier Reef, the coast south of the coral's present distribution lacks the extensive shallow continental shelf required to support large reefs. A few species might find refuge in places like Sydney Harbour, but only a fraction of the diversity of even mobile reef species could exist in such limited spaces.

So what is the prognosis for the world's coral reefs? The complexity of their ecology, and our limited knowledge of key aspects of them, makes the response of the reefs towards warming among the most difficult of climate change outcomes to determine. Nevertheless, the damage already sustained is a strong indication that reefs are sensitive to the perturbations climate change brings, leading me (and many other scientists) to believe that the future for reefs under the emerging new climate is bleak.

Let's imagine what the Great Barrier Reef might look like fifty years

from now. Only fifty of the 400 species of hard coral currently inhabiting
the reef complex are likely to have adapted to using *Symbiodinium* strain
D as partners, and almost all of these heat-hardy species are lumpy rock-
like forms or thick, sturdy types. Not only are such corals relatively
unattractive, but they do not form the labyrinthine structures so necessary
to the reef's biodiversity.[54] It is hard to believe that anything more than a
small proportion of the reef's creatures could survive this transformation.
So, in effect, visitors travelling to Queensland by 2050 may see the Great
Stumpy Reef. Tourism is Australia's second largest income earner, and
the Great Barrier Reef is one of the industry's leading drawcards, so
deciding who will pay to see the wonders of the Great Stumpy Reef is of
more than academic significance. And, with some nations entirely
dependent upon coral reefs for their existence, far more than economics
is at stake.

TWELVE

A WARNING FROM THE GOLDEN TOAD

When the chicha has been drunk, the night grows late and dark, and the fires die down to burning embers, the wisest old man of the tribe tells…of a beautiful miraculous golden frog that dwells in the forests of these mystical mountains. According to the legends, this frog is ever so shy and retiring and can only be found after arduous trials and patient search in the dark woods on fog-shrouded slopes and frigid peaks…The reward for the finder of this marvellous creature is sublime. Anyone who spies the glittering brilliance of the frog is at first astounded by its beauty and overwhelmed with the excitement and joy of discovery; almost simultaneously he may experience great fear. The story continues that any man who finds the legendary frog finds happiness…One story tells of a man who found the frog, captured it, but then let it go because he did not recognise happiness when he had it; another released the frog because he found happiness too painful.

J. Savage, *On the Trail of the Golden Frog*, 1970.

Up to this point in our narrative, not a single species is definitely known to have become extinct because of climate change. In the regions where it is likely to have occurred, such as New Guinea's forests and coral reefs, there's been no biologist on hand to document the event. In contrast, the Monteverde Cloud Forest Preserve in Costa Rica, wherein is situated the Golden Toad Laboratory for Conservation, is blessed with an abundance of researchers. Soon after our fragile planet passed through the climatic magic gate of 1976, abrupt and strange events were observed by the ecologists who spend their lives conducting detailed field studies in these pristine forests.

Although the lion did not lie down with the lamb anywhere in the observable world after 1976, at Monteverde the keel-billed toucan (*Ramphastos sulphuratus*) did nest alongside the resplendent quetzal (*Pharomachrus moccino*)—which in the eyes of rainforest ecologists was as serious a prognostication of impending doom as any biblical omen.

The keel-billed toucan is a lowland bird and its abrupt intrusion into the misty realm of the brilliant red and green quetzal, which is the spiritual protector of the Maya, was a sign of changing conditions high on the mountain. The quetzal can still be seen on Monteverde, but it is not as common as it once was, in part because the keel-billed toucan eats its eggs and young. Some more sensitive bird species have already vanished from the site altogether.

Then, during the winter dry season of 1987, in the mossy rainforests that clothe the mountain's slopes one and a half kilometres above the sea, thirty of the fifty species of frogs known to inhabit the 30-square-kilometre study site vanished. Among them was a spectacular toad the colour of spun gold. The creature lived only on the upper slopes of the mountain, but there it was abundant, and at certain times of the year the brilliant males could be seen by the dozen, gathering around puddles on the forest floor to mate. Aptly named the golden toad (*Bufo periglenes*), its disappearance particularly worried researchers, for it is one of the most spectacular of the region's amphibians and was found nowhere else.

The golden toad was discovered and named in 1966. Only the males are golden; the females are mottled black, yellow and scarlet. For much of the year it's a secretive creature, spending its time underground, in burrows amid the mossy root-masses of the elfin woodland. Then, as the dry season gives way to the wet in April–May, it appears above ground en masse, for just a few days or weeks. With such a short time to reproduce, the males fight with each other for top spot and take every opportunity to mate—even if it's only with a field worker's boot.

In her book *In Search of the Golden Frog* (perhaps 'toad' was too off-putting for a title), amphibian expert Marty Crump tells us what it was like to see the creature in its mating frenzy.[24]

> I trudge uphill…through cloud forest, then through gnarled elfin forest…At the next bend I see one of the most incredible sights I've ever seen. There, congregated around several small pools at the bases of dwarfed, windswept trees, are over one hundred Day-Glo golden orange toads poised like statues, dazzling jewels against the dark brown mud.

On 15 April 1987 Crump made a note in her field diary that was to have historic significance:

> We see a large orange blob with legs flailing in all directions: a writhing mass of toad flesh. Closer examination reveals three males, each struggling to gain access to the female in the middle. Forty-two brilliant orange splotches poised around the pool are unmated males, alert to any movement and ready to pounce. Another fifty-seven unmated males are scattered nearby. In total we find 133 toads in the neighbourhood of this kitchen sink-sized pool.

On 20 April:

> Breeding seems to be over. I found the last female four days ago, and gradually the males have returned to their underground retreats. Every day the ground is drier and the pools contain less water. Today's observations are discouraging. Most of the pools have dried completely, leaving behind desiccated eggs already covered in mold. Unfortunately, the dry weather conditions of El Niño are still affecting this part of Costa Rica.

As if they knew the fate of their eggs, the toads attempted to breed again in May. This was, as far as the world knows, the last great toad orgy ever to occur, and Crump had the privilege to record it. Despite the fact that 43,500 eggs were deposited in the ten pools she studied, only twenty-nine tadpoles survived for longer than a week, for the pools once again quickly dried.

The following year Crump was back at Monteverde for the breeding season, but this time things were different. After a long search, on 21 May she located a single male. By June, and still searching, Crump was worried: 'the forest seems sterile and depressing without the bright orange splashes of colour I've come to associate with this [wet] weather. I don't understand what's happening. Why haven't we found a few hopeful males, checking out the pools in anticipation…?' Yet even after the season closed without another sighting there was no undue pessimism. A year was to pass before, on 15 May 1989, a solitary male was again sighted. As it was sitting just three metres from where Crump made her sighting twelve months earlier, it was almost certainly the same male who, for the second year running held a lonely vigil, waiting for the arrival of his fellows. He was, as far as we know, the last of his species, for the golden toad has not been seen since.

Toads and toucans, it transpired, were just two of the species affected by the changes. Lizards in particular suffered population crashes in the years following 1987, especially the anoles, small relatives of the iguanas; by 1996 two species—the cloud forest anole (*Norops tropidolepis*) and the montane anole (*Norops altae*)—had vanished entirely. Today, the mountain's rainforests continue to be stripped of their jewels, with many reptiles, frogs and other fauna becoming rarer by the year. While still verdant enough to justify its name, the Monteverde Cloud Forest Preserve is beginning to resemble a crown that has lost its brightest and most beautiful gems.

Suspecting that some odd weather event might be the cause of the changes, researchers began to pore over the monthly records of regional temperature and rainfall, but they could find nothing unusual in this data. Fortunately, an alternative and more detailed source of information existed on the mountain top—a weather station is situated on the edge of the study site, and it provided the finer grained record of local changes required to solve the mystery. It would be twelve years before the researchers published their findings, but in 1999 they announced that they had identified the cause of Monteverde's despoliation.

Examination of the meteorological record revealed that, ever since Earth had passed through its first climatic magic gate in 1976, the number of mistless days experienced each dry season had grown, until they had coalesced into runs of mistless days. By the dry season of 1987, the number of consecutive mistless days had passed some critical threshold. It was apparently so subtle as to be undetectable to the researchers working on the mountain, yet it had plunged the entire ecosystem of the mountaintop into crisis. Mist, you see, brought vital moisture, and without it the forest dried out sufficiently to trigger a landslide of catastrophic changes that swept before it mountain birds, anoles, golden toads and other amphibians alike.

Why, the researchers wanted to know, had the mist forsaken Monteverde? The cloud-line is the level at which clouds sit against mountainsides, bringing misty conditions, and beginning in 1976 the bottom of the cloud mass had risen until it was above the level of the forest. The change had been driven by the abrupt rise in sea surface temperatures in the central western Pacific that heralded the magic gate of 1976. A hot ocean had perhaps heated the air, elevating the condensation point for moisture in it. By 1987 the rising cloud-line had, on many days, forsaken the mossy forest altogether and hung about in the sky above, bringing shade but no mist. It was this shade, and the cool it brought, which had been recorded in the original regional weather records and which had first confused the researchers.

The golden toad's permeable skin, and its propensity to wander in daylight hours, had left it exquisitely vulnerable to the desiccation brought on by the run of mistless days. By the time the study was published in 1999, this wondrous creature had been extinct for a decade.

It's always devastating when you witness a species' extinction, for what you are seeing is the dismantling of ecosystems and irreparable genetic loss. The golden toad's extinction, however, was not in vain, for when the explanation of its demise was published in *Nature*, the scientists could make their point without equivocation. The golden toad was the first documented victim of global warming. We had killed it with our profligate use of coal-fired electricity and our oversize cars just as surely as if we had flattened its forest with bulldozers. It was as if, having experienced it, we did not recognise what happiness was.

As the reason for the extinction of the golden toad became thoroughly comprehensible, frog researchers worldwide began to re-evaluate their experiences, for since 1976 many had observed amphibian species vanish before their eyes without being able to determine the cause. Could climate change, they wondered, be responsible?

South Australian Museum researcher Steve Richards has documented a series of amphibian declines in the mossy mountain rainforests of eastern Australia. These began in the late 1970s, when a remarkable creature known as the gastric brooding frog (*Rheobatrachus silus*) disappeared from southeastern Queensland. When first discovered in 1973 this brown, medium-sized frog astonished the world. The surprise came when a researcher looked into a female's open mouth—to observe a miniature frog sitting on her tongue! Not just the frog—scientists were open-mouthed too. This might lead the casual observer to think the species was cannibalistic, but this was not the case; it just had bizarre breeding habits. The female swallows her fertilised eggs, and the tadpoles develop in her stomach until they metamorphose into frogs, which she then regurgitates into the world.

When this novel method of reproduction was announced, some medical doctors understandably got very excited. How, they wondered, did the frog transform its stomach from an acid-filled digesting device into a nursery? They thought the answer might assist in treating a variety of stomach complaints. Alas, they were unable to carry out many experiments, for in 1979—six years after its existence was announced to the world—the gastric brooding frog vanished, and with it went another inhabitant of the same streams, the day frog (*Taudactylus diurnus*). Neither have been seen since.

Australia's gastric brooding frog nurtured its tadpoles in its stomach, which it somehow transformed from an organ of digestion into a brood chamber. The species may well be Australia's first victim of climate change.

Five years after the last gastric brooding frog hopped into oblivion the discovery of another species in the same genus was announced. This one, *Rheobatrachus vitellinus*, lived further north, on Queensland's central coast. It was larger, but otherwise strikingly similar. You may have noticed that it lacked a common name, so it won't be a surprise to learn that the herpetologists' excitement was short-lived. Before it could be studied in detail this species too could no longer be found—its existence as a known species was measured in months rather than years.

In the early 1990s, frogs began to disappear en masse from the rainforests of northern Queensland and, as with the golden toad, these vanishings occurred in otherwise undisturbed rainforest. Today some sixteen frog species (13 per cent of Australia's total amphibian fauna), have experienced dramatic declines. The cause is still debated, but the climate change experienced in eastern Australia over the past few decades cannot have been good for frogs, for a persistence of El Niño-like conditions has brought about a dramatic decline in Australia's east coast rainfall. The latest analyses suggest that at least in the case of the gastric brooder and day frog, climate change was the most likely cause for their disappearance.[30]

When the first global survey of amphibians was completed in 2004 it revealed that almost a third of the world's 6000-odd species was threatened with extinction.[145] Many of these endangered species began their decline after 1976, and according to Simon Stuart of the International Union for the Conservation of Nature, 'There's almost no evidence of recovery.'[145]

After a decade of research, North American scientists produced their own elegant hypothesis that draws together the causes of these declines into a single, unifying concept.[25] Their study focuses on the fate of the amphibians of the northwestern United States, and typical of the patterns they found are those of the western *Bufo boreas*.

Amphibians in the genus *Bufo* are commonly known as toads. One fundamental discovery of the American study was that ultraviolet light retards the development of the toad's embyros, and this in turn makes them more vulnerable to a chytrid-type fungal disease known as *Saprolgenia ferax*, a killer of amphibians worldwide. The toad embryos, it transpires, were receiving more ultraviolet light because their nursery ponds were shallower; this was because persistent El Niño-like conditions since 1976 have brought less winter rain to the Pacific northwest. Even a small change in pond depth can be critical. In ponds 50 centimetres deep just 12 per cent of tadpoles die from the fungus, but when it is only 20 centimetres deep, 80 per cent die.[143] At worst some ponds dried up completely, killing all the tadpoles in them.

To compensate, some frogs tried to breed in larger water bodies, but these contained fish that ate the hatching tadpoles, and between the fungus, the drying ponds and the fish, the region's amphibians had nowhere to go, and so joined the long list of species in free fall towards extinction.

The elegance of this hypothesis lies in drawing together a constellation of impacts under a single dominant factor. In various parts of the world researchers have documented one or more of these changes at work. In the case of the golden toad it was the loss of mist. Australia's frogs have been reported to be infected with chytrid fungus, while elsewhere failing rains or tadpole deaths mean that reproduction is in decline. Yet, whatever the immediate cause, underlying it all is the change in the patterns of our weather brought about by the magic gate year of 1976, and perhaps that of 1998.

THIRTEEN

LIQUID GOLD: CHANGES IN RAINFALL

> Hath the rain a father? Or who hath begotten
> the drops of dew?
> The Book of Job

From the Poles to the equator our Earth spans a range of temperatures from around 40°C below zero to 40°C above, and air at 40°C can hold 470 times as much water vapour as air at −40°C. It's this fact that condemns our Poles to be great frozen deserts, and which dictates that, for every degree of warming we create, our world will experience an average 1 per cent increase in rainfall.[103] But the critical fact here is that this rainfall increase is not evenly distributed in time and space. Instead, rain is appearing at unusual times in some places, and not at all in others. There are even a few favoured places where little change in rainfall is evident.

Over large parts of the world rainfall is increasing, but more rain is not necessarily a good thing for either nature or humans. One of the most certain predictions of climate science is that as our planet warms more rain will fall at high latitudes in winter, and as we have seen, that can be a very bad thing for the inhabitants of the Arctic. Further south, increases in winter rain are also bringing unwelcome change: in 2003 it triggered a deadly avalanche season in Canada, while the British spring of 2004 was so wet that in many regions hay-making was difficult or impossible. Flooding, of course, is expected to increase wherever rainfall does, but as extreme weather events become more common, the incidence of flooding will exceed that brought by a rainfall increase alone.

Here, though, I wish to concentrate on the regions that climate change will tip into perpetual rainfall deficit, for some may be transformed into new Saharas, or at least into regions untenable for human habitation: indeed some already have. A lack of rainfall is often referred to as a 'drought', yet droughts are by their nature transient, and in the areas under discussion there is no prospect that the rain will return. Instead what has occurred is a rapid shift to a new, drier climate.

The first evidence of such a shift emerged in Africa's Sahel region during the 1960s. The area affected was huge—an enormous swathe of sub-Saharan Africa extending from the Atlantic Ocean to Sudan. Four decades have now passed since the sudden decline in rainfall, and there is no sign that the life-giving monsoon rains will return.[166] Even before the decline, the Sahel was a region of marginal rainfall where life was tough. In areas with better soils and more rain, farmers made a living out of their fields, while in the drier wastes camel herders followed their semi-nomadic round in pursuit of feed for their herds. The decreased rainfall has made life difficult for both groups: herders struggle to find grass in what is now a true desert, while the farmers rarely get sufficient rain to stir their fields to life. The world's media periodically shows images of the result—starving camels and desperate families struggling in a dust-filled wasteland.

I remember as a child seeing these images on television, and hearing about how overgrazing and a burgeoning population had caused this human misery. Indeed for decades the West has reassured itself that this disaster was brought on by the people themselves. The argument was that overgrazing by camels, goats and cattle, as well as people gathering firewood, had destroyed the region's thin covering of vegetation, exposing the dark soil and changing the albedo of the area. With constant updrafts of hot, dry air, and no plants to transpire moisture into the atmosphere, the rain-forming clouds had failed to gather, and as this manmade 'drought' lengthened the soil began to blow away. It's an interpretation that has provided opportunities for both environmentalists and moralists to sermonise; but it is wrong in almost every respect.

The true origin of the Sahel disaster was revealed in November 2003, when climatologists at the National Center for Atmospheric Research in Boulder, Colorado, published a painstaking study that used computer models to simulate rainfall regimes in the region between 1930 to 2000.[168] It was a massive exercise, for everything from sea and land temperatures to changes in the region's vegetation needed to be fed into the computer.

In the end the model proved able to simulate past and current climate in the region and revealed that the amount of human-caused land degradation there was far too insignificant to have triggered the dramatic climate shift.[166] Instead, a single climatic variable was responsible for much of the rainfall decline: rising sea-surface temperatures in the Indian Ocean, which resulted from an accumulation of greenhouse gases. The Indian Ocean is the most rapidly warming ocean on Earth, and the computer study showed that as it warms, the conditions that generate the Sahelian monsoon weaken. As a result, by the 1960s the Sahelian 'drought' had begun.

As is commonly the case in such studies, not all of the observed rainfall decline could be explained, which means that some unidentified mechanism was at work. But now, some scientists think that they have found the cause and it's called 'global dimming'.

Global dimming is a phenomenon that cuts down the amount of sunlight reaching Earth's surface. It has caused a cooling of the oceans around Europe, which has further weakened the monsoon. Global dimming is in large part due to particles spewed out into the air by coal-fired power plants, automobiles and factories. This bolsters the argument that the Sahelian catastrophe was not the result of ecological mismanagement by primitive and ignorant pastoralists. As profound as the moral implications of this study are, it seems to have gone all but unnoticed in the world's news media.

In the Darfur region of western Sudan the Sahelian climate shift has driven people to desperation. Camel-herding nomads have been forced to drive their herds onto agricultural lands, where they have come into conflict with farmers. Although the herders are characterised as Arabs, and the farmers as Africans, with the exception of their lifestyles they are culturally and physically indistinguishable. When journalist Tim Judah visited there he was told by the governor of El Fasher that, 'Everyone here is intermarried and lives as one family.'[202] And Judah himself observed that there were just as many frightened and hungry nomads as farmers. With the UN already feeding 1.3 million from a population six times that, the climate change-induced misery looks set to continue.

The Sahelian climate shift is emblematic of the situation faced by the world as a whole, for in it we see the West focusing on religion and politics as the problem, rather than the well-documented and evident environmental catastrophe that is its ultimate cause. For decades we have deluded ourselves about its origins, but the day of reckoning must come. So big is the Sahelian climate shift that it could influence the climate of the entire planet. This was first noted by researchers Joseph Prospero and Peter Lamb, who studied the dust that blows from the Sahel.[167]

Dust is important stuff, because its tiny particles can scatter and absorb light, thereby lowering temperature. These particles also carry nutrients into the ocean and to distant lands, assisting the growth of plants and plankton, and thereby increasing the absorption of CO_2. Around half of the global dust in the air today originates in arid Africa, and the impact of the drying is so great that the planet's atmospheric dust loading has increased by a third. Climatologists are still calculating what will result, but so interconnected is everything to everything else in this Gaian world that a phenomenon of this scale is certain to have an impact.

The citizens of the industrialised world tend to feel that their technology will protect them from Sahelian-scale disasters, but nature has been busy proving them wrong. Australia is a dry country, and Australians—even urban ones—are obsessed with rainfall. The south-western corner of Western Australia once enjoyed one of the most reliable of rainfall regimes. Traditionally the rain fell during the winter, with over 100 centimetres falling annually at some locations. This made the area famed for its primary production, the western wheatbelt being one of the largest and most predicatable centres of grain production in the entire continent. More recently vineyards have spread throughout the wetter areas, and they produce some of the finest and most expensive wines made in the Southern Hemisphere.

Before settlement, most of the southwest was blanketed in a tough, spiny heath-like vegetation known as kwongan, which following the winter rains was transformed into a vast, natural wildflower garden. Only in the tropical rainforest and a similar region in South Africa are more species jammed into a single hectare, and the plethora of plants support ancient, endemic animal families such as the honey-possum (*Tarsipes rostratus*), the salamander-fish (*Lepidogalaxias salamandroides*) and the western swamp tortoise (*Pseudemydura umbrina*). All are adapted to a pattern of winter rain and summer drought. Indeed, over millions of years this is what has shaped them.

During the first 148 years of European habitation of the southwest (1829–1975) the reliable winter rainfall brought prosperity and opportunity. But then things changed and ever since the region has endured a decrease in rainfall averaging 15 per cent. Climate models indicate that about half the decline results from global warming, which has pushed the temperate weather zone southward. The Australian climatologist David Karoly thinks the other half results from destruction of the ozone layer, which has cooled the stratosphere over the Antarctic, thus hastening the circulation of cold air around the Pole and drawing the southern rainfall zone even further southwards.[45]

While a 15 per cent loss may seem trivial, its impact has been considerable. The deficit was felt at once on farms, particularly on the region's margins where variation of a few tens of millimetres makes the difference between a good crop and failure. In these areas wheat is the principal crop, and it's grown in an unusual manner. In the 1960s the goal of the western farmers was to clear a million acres of scrub a year. When the bulldozers had done their work farmers found themselves staring at sterile stretches of sand—some of the most infertile soil to be found anywhere on Earth—for here, as in rainforests, the region's natural wealth was bound up in its vegetation. This, however, was what the farmers wanted, for wheat-growing in the southwest was a gigantic version of hydroponic gardening: farmers drilled in their wheat, dusted the sterile sand with nutrients, then waited for the never-failing winter rains to add water.

By 2004, after decades of nature refusing to 'just add water', the wheat-growing region began retracting westwards, replacing dairying in country once considered too wet for the crop. The Indian Ocean is the ultimate barrier to this succession and, as conditions worsen over the coming century, one high-rainfall activity after another must face being pushed into the sea. But that seemingly trivial 15 per cent reduction in rainfall hides an even greater catastrophe: winter rainfall has in reality declined by more than that, while summer rainfall (which is far more erratic) has increased. Because summer rains cannot be depended upon, farmers do not plant summer crops, so the rain falls on bare fields, allowing the water to soak down to the watertable. There it meets salt, which steady westerly winds have been blowing in from the Indian Ocean for millions of years.

Under every square metre of this land lies an average of between 70 and 120 kilograms of salt. Before land-clearing this didn't matter, for the diverse native vegetation used every drop of water that fell from the heavens, and the salt stayed in its crystalline form. As the summer rains began to fall on the vacant wheatfields, however, water far saltier than the sea began to creep upward, killing everything it touched. The first sign of trouble was a salty taste in the previously sweet brooks of the region. In many cases they quickly became undrinkable, their streamside vegetation died and within a decade or two they had turned into collapsed, salty drains. Today, impoverished and bankrupt farmers are facing the worst case of dry-land salinity in the world. Neither science nor government has been able to provide solutions, and the damage bill is in the billions. Roads, railways, houses and airfields are now besieged by salt, and unless the original vegetation can be returned and induced to grow in the drier and saltier conditions that now prevail, there appears to be no hope of a turnaround.

Western Australia's capital is Perth, a thirsty city of 1.5 million people and the world's most isolated metropolis. There, a taxi driver is likely to be a bankrupt wheat farmer scraping together a living as he tries to sell a now useless farm. For Perth, the most crucial impact from the decline in winter rainfall was less water in the city's catchments, for after 1975 the rain tended to fall in light showers that soaked into the soil and did not reach the dams. Over most of the twentieth century an average of 338 gigalitres of water per year had flowed into the dams that quench the city's thirst. But between 1975 and 1996 the average was only 177 gigalitres—representing a cut of 50 per cent to the city's surface water supply. Between 1997 and 2004 it had fallen to just 120 gigalitres—little more than a third of the flow received three decades earlier.

Severe water restrictions were put in place in 1976, but the situation was soon eased by drawing on a reserve of groundwater known as the

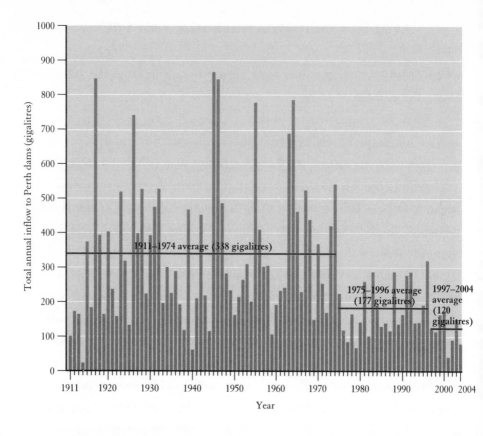

This shows the water flow into Perth's catchments between 1911 and 2004. Large reductions followed the magic gate years 1976 and 1998, and the city has lost two-thirds of its surface water supply over the last thirty years.

Gnangara Mound. For a quarter of a century the city mined this subterranean water, but the failing rains meant that it was not being recharged. In 2001 Perth's dams received virtually no water, and by 2004 the situation of the Gnangara Mound was critical, with the state's Environmental Protection Authority warning that extracting more water from it would threaten some species with extinction.[181] Today, the western swamp tortoise, which is a living fossil, only survives because water is pumped into its habitat.

By early 2005, nearly thirty years after the crisis first emerged, the city's water experts rated the chance of a 'catastrophic failure of supply'—which means no water coming out of the tap—at one in five. Were that to eventuate the city would have no choice but to squeeze what water it could out of the Gnangara Mound, and in doing so destroy much ancient and wondrous biodiversity, and *still* the fix would only be temporary. Plans have now been laid for a desalination plant at a cost of $350 million dollars, making it one of the largest in the Southern Hemisphere. This though, would still supply only around 15 per cent of the city's water.

Australia's east coast is no stranger to drought, but the dry spell that began in 1998 is different from anything that has gone before. Thus far it has consisted of seven years of below average rainfall; and this is a 'hot drought', with temperatures around 1.7 degrees warmer than in previous droughts, making it of exceptional hostility.[169] The cause of the decline of rainfall on Australia's east coast is thought to be a climate-change double whammy—the loss of winter rainfall and the prolongation of El Niño-like conditions.[134]

The resulting water crisis here is potentially even more damaging than the one in the west, for cities such as Sydney lack the groundwater resources enjoyed by Perth. Its only buffer against rainfall deficits is its dams, meaning that a decline in stream flow translates at once into water hardship. Sydney's water supply is one of the largest domestic water supplies in the world, able to store four times as much per capita as New York's water supplies, and nine times as much as London's. Yet even this capacious storage has proved insufficient. Between 1990 and 1996 the total flow into all eleven of Sydney's dams had averaged 71,635 megalitres, but by 2003 this had dropped to just 39,881 megalitres, a decline of 45 per cent. As of mid 2005 the situation remains critical, with Sydney's 4 million residents having around two years' worth of supply in storage.[203] If the dry conditions continue, this leaves very little time to arrange for alternative water sources such as large-scale desalination plants.

Across the Pacific Ocean, much of the American west is in its fifth year of drought. Research shows that such dry conditions have not been seen in the region for around 700 years, during which time the American southwest was even warmer than it is today.[171] This suggests a relationship between drought and warmer conditions, and as with the Sahel the link seems to lie in rising ocean temperatures.[44]

Between 1998 and 2002 the Pacific Ocean was in an unusual state. Waters in the eastern tropical Pacific were a few degrees cooler than normal, while those in the central western Pacific were far warmer— around 30°C—than average. These conditions shifted the Jet Stream northwards, pushing storms that would usually track at around 35° of latitude to north of 40°. 'This is a reinforcement of the connectedness of the climate system over great distances and long time scales,' observed Kelly Redmond of Nevada's Desert Research Institute. And, of course, what was driving warmer ocean temperatures was CO_2 in the atmosphere.

The drought conditions in the American west are frequently portrayed in the media as being part of a natural cycle. The only way to be absolutely sure if this is the case is to wait for the decades or hundreds of years required for any natural cycle to play itself out. But the fact that the changes are consistent with those expected to result from global warming, and that they have been observed during warm times in the past, is worrying. Furthermore, the potential of climate change to spawn drought almost anywhere on the planet is so great that leading climatologists have recently warned that 'it would be a mistake to assume any region is safe from megadrought'.[231] In this regard, it is worth pointing out that the near record rains the US experienced over the winter of 2005 in parts of the southwest were not sufficient to make up for the preceding dry years, while the northwest remains in the grip of unprecedented drought.

Much of the water in the American southwest comes in the form of winter snow that accumulates on its high mountains. Because it melts over the spring and summer, it provides stream flow when most needed by farmers. In effect the snowpack has offered an inexpensive form of water storage that has minimised the need for dams. The amount of snow that falls has always varied considerably from year to year, and this can hide any longer-term trend from the casual observer. Over the last fifty years, however, there has been a decline in the average amount of snow received. If this trend continues for another five decades, western snowpacks will reduce by up to 60 per cent in some regions, which could cut summertime stream flow in half.[165] This will devastate not just water supplies, but hydropower and fish habitats as well.

Changes in the overall volume of snowfall, however, are not nearly as worrying as changes in the way the snowpack forms and melts. Over the past fifty years, the southwest region has warmed by 0.8°C—slightly more than the global average and, even in regions that are now receiving more snow, this and seasonal changes in rainfall and temperature are affecting water supply. These factors have conspired to reduce the snowpack. This is because the higher temperatures are melting it before it can consolidate. On the whole the snowpack is melting earlier, which means that the peak of runoff into streams is now occurring three weeks sooner than in 1948. This leaves less water for the height of summer, when it's most needed, but increases water flow in winter and spring, which may lead to more flooding. With temperatures in the region set to rise between 2°C and 7°C over this century (unless we significantly reduce CO_2 emissions), it can be anticipated that most streams will eventually flow in winter, when the water is least needed.[165]

I can imagine the response of many people to reading this: 'So what? We'll just build more dams.' And it is possible that, as the crisis deepens, this is what people will do. But there are a limited number of sites suitable for dams in the region, and dams mean that farmers will pay for water storage that was once provided by nature. Besides, the changes under way are so vast that even a new program of dam-building is insufficient to counter them. Researchers forecast that snowpack changes could lower farm values by 15 per cent, costing billions. The biggest problem, however, is surely to do with the cities of the US west, which are tethered to ever dwindling water supplies.

These vast metropolises are impossible to relocate and some—as it was with the ancient cities of Mesopotamia—may, if the rate of change accelerates, have to be abandoned. If this sounds extreme, it's well to remember that we are only at the beginning of the West's water crisis. Five thousand years ago, when the American southwest was a little warmer and drier even than it is today, the Indian cultures that had flourished across the region all but vanished. Only when conditions cooled again was the region habitable. For more than a millennium the southwest was little more than one big ghost town.[227]

FOURTEEN

AN ENERGETIC ONION SKIN

> Some…storms are so violent that no human
> structures can resist them, while the largest and most vigor-
> ous trees are torn to pieces or overturned by them. If our
> atmosphere [received] a somewhat greater amount of sun
> heat…these tempests might be so increased in frequency and
> violence as to render considerable portions of the globe
> uninhabitable.
>
> Alfred Russel Wallace, *Man's Place in the
> Universe*, 1903.

In 2003 climate scientists announced that over just a few years the tropopause had risen by several hundred metres. Why should this small
adjustment between layers of the atmosphere occurring eleven kilometres
above our heads worry us? For the very good reason that climatologists
now realise that the tropopause is where much of our weather is generated. Change it, and you alter not only weather patterns but extreme
weather events as well.

The cause of the change is a dyad of human-caused pollution—ozone-destroying chemicals and greenhouse gases. As we have seen, chlorofluorocarbons (CFCs) destroy ozone, and ozone absorbs ultraviolet radiation, in the process emitting heat. With less heat produced in the stratosphere, this layer of the atmosphere has cooled and shrunk. Meanwhile in the troposphere, ever-increasing levels of greenhouse gases are trapping more heat, causing it to expand. Between these two effects, the tropopause has been rapidly ascending. At the same time, changes within the troposphere itself have been manifesting their own effects. By warming the troposphere we both change the weather patterns globally and increase the likelihood of extreme weather events.

As the troposphere has warmed over the past decade, the world has seen the most powerful El Niño ever recorded (1997–98), the most devastating hurricane in 200 years (Mitch, 1998), the hottest European summer on record (2003), the first South Atlantic hurricane ever (2002), and one of the worst storm seasons ever experienced in Florida (2004). This series of events, many would argue, indicates that the potential for the new climate to generate extremes is already increasing.

Where do you think the energy to power a hurricane comes from? 'A hurricane,' Frederick Lutgens and Edward Tarbuck tell us in their textbook of atmospheric studies, 'is a heat engine that is fuelled by the latent heat liberated when huge quantities of water vapour condense. To get this engine started, a large quantity of warm, moist air is required, and a continuous supply is needed to keep it going.'[117] We're all familiar with the principle that evaporation can carry heat into the atmosphere: on a hot day we all perspire, and as our sweat evaporates it carries heat from our body into the air. It's a highly effective form of heat transfer, for the evaporation of just one gram of water from our skin is sufficient to transfer 580 calories of heat.[199] Think of the difference in scale of your body and an entire ocean, and you can sense the power that heat energy derived from evaporation carries into the great aerial ocean.

It's not widely appreciated just how much extra latent heat the hot air engendered by climate change can carry. For every 10°C increase in its temperature, the amount of water vapour that the air can hold doubles; thus air at 30°C can hold four times as much 'hurricane fuel' as air at 10°C.[117]

Perhaps the most marked change in hurricanes since around 1950— when global warming began to be felt—is a change in their tracks. One of the best-documented examples of this comes from eastern Asia. The frequency of typhoons ravaging the East China and Philippine seas has decreased since 1976, but the number in the South China Sea has increased.[105] Further westward, in the Arabian Sea and in the Bay of Bengal, there have been fewer typhoons, which is good news for the millions living near sea level in these regions. Another very marked change has been noted at high latitudes in the Southern Hemisphere, where there has been a dramatic decrease in the number of cyclones occurring over the sub-Antarctic Ocean south of latitude 40, but a modest increase in the Antarctic Ocean.[106] Although the time frame of the change is short, the last couple of decades have also seen intense low pressure systems develop between 30° and 40°S in parts of the Southern Hemisphere, with one extraordinary low-pressure system approaching cyclone intensity in February 2005.

There are disturbing signs that hurricanes are becoming more frequent in North America. In 1996, 1997 and 1999 the United States endured more than twice the number of hurricanes experienced annually during the twentieth century, and what 1998's hurricanes lacked in numbers, they more than made up for in intensity. Hurricane Mitch tore through the Caribbean in October of that year, killing 10,000 people and making 3 million homeless. With wind speeds of up to 290 kilometres per hour, Mitch was the fourth strongest Atlantic Basin hurricane ever recorded. It was also the most damaging storm to hit the Americas in 200 years; only the Great Hurricane of 1780 which killed at least 22,000 people was more severe in its impact.

After a few years of relative calm the storms returned with a vengeance in 2004, when four major tropical storms crossed the Florida coast in quick succession, devastating large parts of the state. Many of the homes damaged by these storms are still uninhabitable, and the US Weather Bureau predicts that the 2005 hurricane season is likely to be more destructive than usual. The season may of course pass serenely, but with hurricane fuel increasing in the atmosphere, it is only a matter of time before the storms return with redoubled fury. Given the extent of the damage wrought in 2004, a recurrence of intense hurricanes within the next few years could send Floridan real estate values plummeting.

In the wake of hurricanes come floods, and because warmer air is able to hold more water vapour, the incidence of severe floods is rising and expected to rise further. In the summer of 2002 two-fifths of the Republic of Korea's annual rainfall fell in a week, wreaking such destruction that the nation had to mobilise its troops to help flood victims. At the same time China suffered floods of historic magnitude, with 100 million people affected.

Looked at globally, the increase in flood damage over recent decades has been profound. In the 1960s around 7 million people were affected by flooding annually. Today that figure stands at 150 million.[112] And in the wake of floods come plagues. Cholera breeds in the stagnant and polluted water, and mosquitoes that can spread malaria, yellow fever, dengue fever and encephalitis proliferate. Even plague can benefit from the disturbance as fleas, rats and humans are brought into close proximity as they crowd together on higher ground.

Because extreme weather events by their very nature are rare, it can be a long time before sufficient data accumulates to detect a trend. Less extreme changes in temperature and rainfall are a lot easier to quantify, and with climate records going back centuries, Europe is a great place to start looking for these impacts. The 1990s were the warmest decade in central England since records began in the 1660s, with 1998 the warmest year ever and 2001 the third warmest. As a result, the growing season for plants has been extended by a month, heatwaves have become more frequent and winters much wetter, with heavier rain.[104] The Hadley Centre is a world leading institution set up to predict and examine climate change impacts. Situated in Exeter, England, it has determined that the United Kingdom has experienced a significant increase in severe winter storms, a trend that is predicted to continue.[103]

On the continent more alarming events have occurred. The European summer of 2003 was so hot that, statistically speaking, such an outlandish event should occur no more often than every 46,000 years.[52] It was worsened by water stress to plants, which restricted their emissions of moisture. With less of the Sun's heat used up in evaporation, more of it warmed the air. The heatwave was so extreme that 26,000 people died during June and July, when temperatures exceeded 40°C across much of the continent. Heatwaves, incidentally, kill a large number of people worldwide each year; even in the climatically turbulent US, heat-related deaths exceed those from all other weather-related causes combined.[117] And just one year after the European heatwave, Egypt experienced one of the highest temperatures ever recorded: 51°C.[162]

Other broad climatic studies have been completed in the US and Australia. In 2003 climatologists published a detailed study of a century's worth of climate records from across North America. They concentrated on changes in temperature, because this is the most direct indicator of climate change, and discovered that prior to 1950 there was no detectable influence on North America's climate from human activities. After 1950, however, the story was very different, for they found abundant evidence that the burning of fossil fuels had not only caused a mean increase in temperature, but that it had decreased the temperature gradient from north to south, altered the temperature contrast between land and sea, and reduced the daytime temperature range. In short, this conservative study—which did not attempt to examine the incidence of extreme weather events or changes in rainfall—established beyond reasonable doubt the fact that climate change is having an impact on the North American continent.[21]

In terms of extreme weather events, it's worth recording that the United States already has the most varied weather of any country on Earth, with more intense and damaging tornadoes, flash floods, intense thunderstorms, hurricanes and blizzards than anywhere else.[117] With the intensity of such events projected to increase as our planet warms, in purely human terms the United States would seem to have more to lose from climate change than any other large nation. Indeed, its ever-spiralling insurance bill resulting from severe weather events and its growing water shortages in the west mean that the US is already paying dearly for its CO_2 emissions.

As we have seen from its abrupt decreases in rainfall, Australia too is suffering the effects of climate change. Many other impacts, however, have been documented, including an increase in the number of very hot days, an increase in night-time temperatures, a decrease in very cold days and a decrease in the incidence of frosts.[4] Some regions, such as around Alice Springs in central Australia, have experienced an increase in temperature of more than 3°C over the twentieth century. There has also been an increase in the occurrence of intense cyclones, as well as severe low-pressure systems in southeastern Australia, particularly over the last twenty years. The frequency of floods has also increased, particularly since the 1960s.[160] Taking a broad view, it's difficult to find two nations that have been more severely disadvantaged by climate change than the US and Australia.

Some regions of the world, in contrast, have so far recorded little change. India in particular seems to be an exception to this picture of ever-rising discomfort, for the subcontinent has as yet been little-affected. Indeed what news we have appears to be good, for apart from Gujarat and western Orissa, most of the region is experiencing less drought than twenty-five years ago and, as we have seen, cyclones seem to be bypassing the Bay of Bengal. Extreme temperatures also seem to be less frequent over most of northern India than in times past, though they are becoming more common in the south. Only northwestern India is experiencing a marked increase in extremely hot days, and the heatwaves there are exacting a heavy toll in human lives.[107]

It has not been my intention here to provide commentary on the climate of every region of the world, only to demonstrate the types of changes in weather that have been documented to date in response to the 0.63°C increase in temperature. There is one impact which results from global warming, however, that is unobtrusively manifesting itself and which is being felt on all continents approximately equally: all of them are shrinking. This is because, courtesy of heat and melting ice, the oceans are expanding.

Is this a threat to humanity? Let's see how far the waters will rise, and at what speed.

FIFTEEN

PLAYING AT CANUTE

> The evening came, the rider of the storm sent down the rain. I looked out at the weather and it was terrible…With the first light of dawn a black cloud came from the horizon; it thundered within where Adad, lord of the storm, was riding…Then the Gods of the Abyss rose up; Nergal pulled out the dams of the nether waters, Ninurta the warlord threw down the dykes, and…the God of the storm turned daylight into darkness.
>
> *The Epic of Gilgamesh*

Nestled deep within the human psyche lies a primal fear of the awful power of water. The epic of Gilgamesh tells of it, as does Noah's flood and hundreds of lesser-known myths from around the world. As we have seen, our species' cradle was most likely the lake region of the African rift valley, where our ancestors foraged on the bounty of fish, shellfish, birds and mammals. We have sought to live close to water ever since, for water draws living things from near and far. Camp near a waterhole and sooner or later animals will come to drink. For deeply embedded reasons, our species has always preferred to live with a water view, especially if it includes a beach, a lake or a lawn cropped short as if by great grazing beasts. Real estate agents well know our housing preferences and the amount we are willing to pay for them. Today, two out of every three people on Earth live within eighty kilometres of the coast, and yet in our subconscious we understand that the waters can rise over the land, making all of our hard-won real estate count for nought.[124]

Fifteen thousand years ago the oceans stood at least 100 metres lower than they do today. Then, the North American continent was a veritable empire of ice, exceeding even the Antarctic in the volume of frozen water it supported. As the great American ice caps melted they alone released enough water to raise global sea levels by 74 metres. The sea rose rapidly until around 8000 years ago, when it reached its present level and conditions stabilised. All around the world people watched the waters rise, at times so fast as to change the coastline from year to year. Today, even a modest sea level rise would be disastrous, for the human population along coastlines is dense, and many of us lead vulnerable lives.

Although it is not related to climate change, the catastrophic Asian tsunami of 2004 gives some indication of how devastating rising seas and turbulent weather might be. The Netherlands is already planning for the construction of a super-dyke to save it from the encroaching ocean, and the Thames barrier is to be strengthened. But countless millions of others live beside the sea—some in expensive real estate, others in humble villages—who have no protection. In Bangladesh alone, more than 10 million people live within one metre of sea level.[119]

All that remains of the great Northern Hemisphere ice caps today is the Greenland ice sheet, the sea ice of the Arctic Ocean and a few continental glaciers, and there are signs that after 8000 years these remnants are beginning to melt away. Alaska's spectacular Columbia Glacier has retreated twelve kilometres over the last twenty years; while in a few decades' time there will be no glaciers left in America's Glacier National

Park. Glaciers such as these only contain enough water to alter the sea level by a matter of centimetres. The Greenland ice cap, however, is a true remnant of the continental ice domes of the type mammoths would recognise, and it contains enough water to raise sea levels globally by around seven metres. In the summer of 2002, it, along with the Arctic ice cap, shrank by a record 1 million square kilometres—the largest decrease ever recorded.[155] Two years later, in 2004, it was discovered that Greenland's glaciers were melting ten times faster than previously thought.

So you might be surprised to learn that temperatures remain cold—indeed they are cooling—over the highest parts of both the Greenland and Antarctic ice domes. These are the only places on Earth where significant negative temperature trends are occurring. This is comforting, for a recent study has concluded that should the Greenland ice cap ever melt it would be impossible to regenerate it, even if our planet's atmospheric CO_2 was returned to pre-industrial levels.[147]

The greatest extent of ice in the Northern Hemisphere is the sea ice covering the polar sea, and since 1979 its extent in summer has contracted by 20 per cent. Furthermore the remaining ice has greatly thinned. Measurements taken using submarines reveal that it is only 60 per cent as thick as it was four decades earlier. This prodigious melting, however, has no direct consequence for rising seas, any more than the melting ice cube in a glass of scotch raises the level of liquid in the glass. This is because the Arctic ice cap is sea ice, nine-tenths of which is submerged, and when it melts it condenses into water in precisely the same proportion as it projected from the sea. Only land ice, as it melts and runs into the sea, adds to sea levels.

Although the melting of sea ice has no direct effect, its indirect effects are important. At its current rate of decline, little if any of the Arctic ice cap will be left by the end of this century, and this will significantly change the Earth's albedo. Remember, one-third of the Sun's rays falling on Earth is reflected back to space. Ice, particularly at the Poles, is responsible for a lot of that albedo, for it reflects back into space up to 90 per cent of the sunlight hitting it. Water, in contrast, is a poor reflector. When the Sun is overhead it reflects a mere 5 to 10 per cent of light back to space, though, as you may have noticed while watching a sunset by the sea, the amount of light reflected off water increases as the Sun approaches the horizon. Replacing Arctic ice with a dark ocean will result in a lot more of the Sun's rays being absorbed at Earth's surface and re-radiated as heat, creating local warming which, in a classic example of a positive feedback loop, will hasten the melting of the remaining continental ice.

As recently as 2001, rising seas looked to be one of the least pressing problems confronting humanity as a consquence of climate change, for over the preceding 150 years the oceans had only risen by 10 to 20 centi-metres, which amounts to 15 millimetres per year—around a tenth as fast as your hair grows.[70] Over the last decade of the twentieth century, however, the rate of sea level rise doubled to around three millimetres per year. While still only a fifth as fast as your hair grows, scientists are concerned at the momentum of the rise, for the sea is the greatest jugger-naut on our planet, and when movements within it reach a certain pace, all the effort of all the people on Earth can do nothing to slow it.

The oceans, of course, are massive when compared with the atmos-phere, having 500 times the mass, and they are very dense. So when we think of the atmosphere changing the oceans, we must imagine something like a VW Beetle pushing a tank down a slope. It takes effort to get the monster moving, but when it does shift there's little the Beetle can do to alter the tank's trajectory. Another factor important in slowing the oceans' reaction to climate change is the stratification of its waters. If all the oceans' water was homogenised to one temperature, it would be a chilly 3.5°C. Away from the Poles the oceans' upper layers are far warmer, but they become ever cooler until, in the depths (because the water is salty), the temperature can be below freezing point.

Any cooling of the surface helps the water layers to mix, thus speeding the cooling process. As the oceans warm they become more stratified, and as a result water mixing from the surface to the depths is impeded so that it takes a long time for the heat to finds its way to the abyssal plain kilometres below.[64] This means that when Earth is on a cooling trend, there is little lag between the reduction of greenhouse gases and the changed climate they entail. When our planet is heating, however, it takes the surface layers of the oceans about three decades to absorb heat from the atmosphere, and a thousand years or more for this heat to reach the ocean depths; all of which means that—from the perspective of global warming—the oceans are still living in the 1970s.

Despite this great inertia, rises in temperature are occurring at the surface of the oceans, and information is also emerging for a sharp rise in temperature at depth.[55] There is nothing we can do to prevent this slow transfer of heat from air to sea, which is very bad news, for the heat acts in two ways to cause rising of the waters.

When most of us think of a rising sea, we imagine melting glaciers and ice caps pouring into the oceans. Over the past century, however, much of the sea level rise has come from an expansion of the oceans, for warm water occupies more space than cold. This 'thermal expansion' of the oceans is expected to raise sea levels by 0.5 to 2 metres over the next 500 years. In 2001 the Intergovernmental Panel on Climate Change estimated that (in round figures) the oceans would rise a mere 10 centimetres to a metre over this century. Thermal expansion, they suggested, would contribute between 10 and 43 centimetres of this rise, while melting mountain glaciers would add up to 23 centimetres more, mostly from the melting of non-polar glaciers and from Greenland.

In the late 1990s, when the panel was compiling its report, the rate of melting of many glaciers was not known, and the situation around the South Pole was particularly uncertain. Heroic scientific efforts have now yielded new data, making the science of sea level change one of the fastest moving aspects of climate science. Typical of this new generation of studies is one published by Eric Rignot of the Jet Propulsion Laboratory, Pasadena, and his collaborators.[153] They measured the melt-rate of the

Patagonian ice fields—the largest temperate ice masses in the Southern Hemisphere—and found that they are contributing more water per unit area to global sea level rise (0.1 mm per year) than even the gigantic glaciers of Alaska.

But it's Antarctica that provides the most alarming news of melting ice. By 2004 one scientific report after another was crowding the pages of learned journals with news of ominous changes to the ice of the Antarctic Peninsula and adjacent areas. These studies make clear that a great domino effect—wherein the destabilisation of one ice field leads to the destruction of a neighbour—is playing itself out at the southern extremity of the world. Because the decline is affecting ever-larger expanses of ice, it is becoming evident that melting polar ice will be by far the greatest contributor to a rising sea in coming decades.

The first dramatic indications that all was not well came in February 2002 when the Larsen B ice-shelf—at 3250 square kilometres it was the size of Luxembourg—broke up over a matter of weeks. Although scientists knew that the Antarctic Peninsula was warming more rapidly than almost anywhere else on Earth, the speed and abruptness of Larsen B's collapse shocked many. In the aftermath, scientists learned that there was an important and hitherto overlooked exception to the rule that melting of sea ice does not affect sea levels. Almost immediately after the break-up, the glaciers that fed into the now fragmented ice sheet began to flow more rapidly. Glaciers, of course, flow much more slowly than rivers. Yet they do flow, and the collapse of Larsen B demonstrated forcefully that one of the most important features determining glaciers' speed is the extent of ice at their mouth. A thick ice sheet acts much like a dam, slowing the flow of the glacial ice to the sea, thereby restricting its rate of melting. Remove the ice plain and the glacier speeds up.

It's difficult and expensive to study Antarctica's glaciers and ice sheets, but the fate of the Larsen B soon had researchers looking both at the details of its demise and at other ice-shelves in the region. In 2003 a study summarising a decade of satellite data revealed the ultimate cause of

Larsen's collapse. Summer melting at both the top and the bottom of the ice sheet, brought about by warming of both the atmosphere and the ocean, had so thinned it and riven it with crevasses that its destruction was inevitable.[148] But melting of the ice from below was the most important factor. While the Weddell Sea's deep waters, which flow past the ice, were still cold enough to kill a person in minutes, they had warmed by 0.32°C since 1972, and this change was enough to initiate the melting.[148]

Scientists are convinced that sometime this century the rest of the Larsen ice-shelf will break up, but by then our attention will be gripped by the fate of far greater ice-masses.[152] The first to enter our consciousness is likely to be the Amundsen ice plain, an extensive area of sea ice off the coast of West Antarctica. In late 2002 a team of scientists led by NASA researchers discovered that it was thinning rapidly. In their study, published in October 2004, they reported that large sections of the ice plain had become so thin that they were nearing a point which could allow them to float free of their 'anchors' on the ocean bed and collapse like Larsen B.[72] The fatal moment for the Amundsen, they ventured, could be as little as five years off, for already its thinning had led to a quickening of glacial flow. At the time of the survey, the glaciers feeding into the Amundsen had increased their rate of discharge to around 250 cubic kilometres of ice per year—enough to raise sea levels globally by 0.25 of a millimetre per annum. As there is enough ice in the glaciers feeding into the Amundsen Sea to raise global sea levels by 1.3 metres, their increasing rate of flow, and the incipient break-up of their ice-plain 'brake', are of concern to everyone.

Across the Antarctic Peninsula lies one of the world's largest surviving expanses of sea ice. The West Antarctic ice sheet is also tenuously anchored to the bottom of a shallow sea. The possibility that it may destabilise was raised back in the 1970s, when University of Ohio glaciologist John Mercer pointed out the similarities between it and the Eurasian Arctic. Both regions, he noted, are shallow seas of similar topography that do (or did) support vast ice sheets. The ice sheets of the

Eurasian Arctic broke up in spectacular fashion between 15,000 and 12,000 years ago and Mercer worried that, as a result of global warming (something which was then almost unheard of), the West Antarctic ice sheet may soon do the same thing.[182]

It was recently discovered that the West Antarctic ice sheet is bounded by rapidly moving 'ice streams' that flow over gravels which, in certain circumstances, facilitate their flow.[154] Just how difficult it is to measure the rate of flow of these 'streams' was demonstrated by a two-week study of the Whillans ice stream. It was long thought to be stable—indeed it was thought to be slowing in its rate of flow—which would have been a good sign for stability of the ice sheet overall. Yet the study revealed that it could move at the extremely rapid rate—for ice, at least—of one metre per hour! This, however, only occurred when the tide was right; when it wasn't, the ice stream stopped.[138] With the ice stream so finely balanced, it's easy to see how rising sea levels or thinning ice might make rapid flow permanent.

If the West Antarctic ice sheet ever does detach itself from the sea floor, it would add 16 to 50 centimetres of sea level rise by 2100. Even worse, the glaciers feeding into it would accelerate, adding much more to sea levels. In all, the 3.8 million cubic kilometres of sea and glacial ice contained and held back by the West Antarctic ice sheet comprise enough water to raise global sea levels by six to seven metres.

There is one bright spot in all of this. The increased precipitation occurring at the Poles is expected to bring more snow to the high Antarctic ice cap, which may compensate for some of the ice being lost at the continent's margins, though just how much compensation this will bring, and for how long, is currently unknown.

So swift have been the changes in ice plain science, and so great is the inertia of the oceanic juggernaut, that climate scientists are now debating whether humans have already tripped the switch that will create an ice-free Earth. If so, we have already committed our planet and ourselves to a rise in the level of the sea of around sixty-seven metres. The next great question would be, how long will it take for the ice to melt? Many scientists think that, regardless of the amount of melting in store, the bulk of the sea level rises will occur after 2050, and it will take millennia for all of the ice to melt. Still, some scientists are predicting a rise in sea levels of three to six metres over a century or two.[231]

Predicting the future has never been humanity's strong suit, but with technological advances made over two decades—including satellite surveillance data of changes at the surface of our planet, better computers, and a firm grasp of Earth systems such as the carbon cycle—scientists have been able to build virtual worlds to see the approximate shape of things to come, and how things might stand if we change our ways. These wondrous new playthings of science have much to tell us about our climatic future over coming decades.

3

THE SCIENCE OF PREDICTION

SIXTEEN

MODEL WORLDS

We have made the natural world our laboratory, but the experiment is inadvertent and thus not designed to yield easily decipherable results…There are unsettling indications that…models are underestimating rather than overestimating the climatic consequences of greenhouse gas build-up.

Lee Kump, 'Reducing Uncertainty about Carbon Dioxide As a Climate Driver', NASA, 2002.

The science of predicting the impact of global warming on Earth's climate has its origins in weather forecasting. Under the leadership of Captain Fitzroy (of Darwin and the *Beagle* fame), the British Meteorological Service was one of the first institutions to develop a scientifically based weather prediction service. Today, with the establishment of the United Nations World Meteorological Organization, scientific activity relating to climate and weather is co-ordinated on a global scale. One hundred and eighty-five countries participate in this program, and between them they monitor 10,000 land-based observation stations, 7000 ship-based ones and ten satellites.

The basic tool used in climate change prediction is a computer model of Earth's surface and the processes at work there. Scientists then vary the inputs, allowing them to see, for example, how our climate might respond to a doubling of CO_2 in the atmosphere, or how the ozone hole affects climate. Early models were restricted to examining circulation patterns in the atmosphere. Hence, all such models—even sophisticated ones that simulate everything from the carbon cycle to vegetation—are known as 'global circulation models'. Little more than fifty years ago the most sophisticated model of Earth's atmospheric circulation was a dishpan full of water on a turntable, which, as it rotated in a laboratory at the University of Chicago, had its rim warmed by a flame representing the tropic Sun.[70] As primitive as it was, the model displayed the currents in the same relative position as the Roaring Forties are in the real world. And to the surprise of the dishpan's creator it even produced a model Jet Stream and eddies resembling storms. Buoyed by the success of their experiment, researchers had by 1949 turned to computers to simulate the atmosphere.

As early as 1975, Syukuro Manabe, who was then working at the US Weather Bureau, and his collaborator Richard Weatherald, used computer models to investigate the consequences of a doubling of CO_2 in the atmosphere.[88] They found that it would cause a rise in the average surface temperature of the Earth of 2.4°C. By 1979 more technologically advanced models had been employed, and these suggested that the rise was more likely to be 3.5 to 3.9°C, give or take a couple of degrees.[70] Astonishingly, for over twenty years this prediction and its degree of uncertainty hardly changed: in 2001 the Intergovernmental Panel on Climate Change (IPCC) was still giving the outcome as 3°C, give or take a couple of degrees. The explanation seems to be that while the increasingly sophisticated computer models eliminated sources of uncertainty in their programs, they had to incorporate more uncertainty from the real world. This situation, however, is now changing.

Today there are around ten different global circulation models seeking to simulate the way the atmosphere behaves, and to predict how it will behave in future.[23] The most sophisticated of them reside at the Hadley Centre in England, the Lawrence Livermore National Laboratory in California, and the Max Planck Institute for Meteorology in Germany. Though all three centres are able to reproduce the general trends in temperature the Earth experienced over the twentieth century, an independent audit rates the Hadley Centre as the world leader.[70]

The Hadley Centre for Climate Prediction and Research has the appearance of the modern cathedral of climate change research. The new building, completed in late 2003, soars overhead, an elegant amalgam of glass and steel designed to minimise energy use and impact on the environment. In this complex more than 120 researchers strive to reduce the uncertainty of predictions by producing ever more sophisticated models that mimic the real world.

If our planet was a uniform black sphere, the Hadley people would have an easy task, for a doubling the CO_2 in the atmosphere would then raise the surface temperature of our hypothetical sooty sphere by 1°C. But the Earth is not black, nor is its surface uniform. Instead it is bumpy, and blue, red, green and white; and it is the white parts—much of which is cloud—that are giving the researchers headaches. Clouds cloud the issue, so to speak, because no one has yet developed a theory of cloud formation and dissipation; and because clouds can both trap heat and reflect sunlight back into space, they can, according to circumstances, powerfully heat or cool.

So how good is the Hadley Centre's cloudy, computerised crystal ball at predicting Earth's future? There are four major tests that any global circulation model must pass before its predictions can be deemed credible. The first is whether its physical basis is consistent with the laws of physics—the conservation of mass, heat, moisture and so on. Second, can it accurately simulate the present climate? Third, can it simulate the day to day evolution of the weather systems that make up our climate? And finally, can the model simulate what is known of past climates?

Computer models such as those run at the Hadley Centre pass all of these tests with a reasonable degree of accuracy, yet new discoveries in the real world are constantly forcing changes on the computer models.[89]

The Canadian researcher Nathan Gillett and his colleagues have recently documented how human-induced climate change is altering sea level pressure. This is the first clear evidence of greenhouse gases directly affecting a meteorological factor other than temperature.[43] Increases in sea level pressure had not previously been incorporated in global circulation models, leading to an underestimate of the impact of climate change on storms in the north Atlantic.

Sceptics who continue to deride the global circulation models include Jack Hollander, emeritus professor of Energy and Resources at the University of California. In his most recent book, *The Real Environmental Crisis*, Hollander opines that 'computer simulations…do not provide an adequate basis for the catastrophic generalisations about future climate… In any case, for most of us it is difficult to distinguish between solid empirical evidence and speculation based on highly uncertain computer models.'[108]

Hollander's division of empirical evidence from speculation reveals a poor understanding of how computer models work. All models draw from evidence and incorporate as much empirical data as possible to build testable hypotheses about future change. As long as scepticism is based on a sound understanding of science it is invaluable, for that is how science progresses. But poor criticism can lead those who are unfamiliar with the science involved into doubting everything about climate change predictions.

One of the most abused aspects of climate change science is the discrepancy between temperature measurements provided by the World Meteorological Organization's 17,000 thermometers (which are housed

(A)

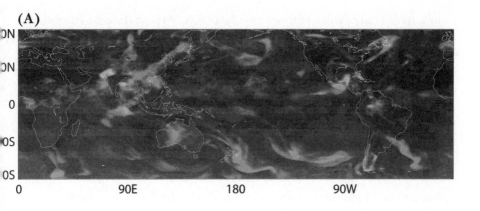

The weather for 1 July 1998. (A) is the Hadley Centre's computer simulation of world weather for the day; (B) is the actual weather as observed by satellite. The white arrows indicate cloud area that the computer failed to simulate, but otherwise the two images are very similar.

(B)

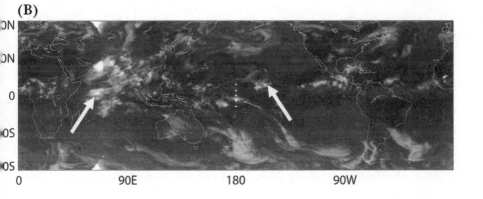

in louvred cubes called Stevenson's boxes), and those derived from their ten satellites. The thermometers were providing seemingly incontrovertible evidence that our planet's surface was warming at the rate of 0.17°C per decade—which, incidentally was the rate predicted by the computer models—yet the satellites indicated a much smaller rate of warming for the lower atmosphere. This was a gift to those seeking to dismiss notions of climate change. Nonetheless they had to twist the data, for to argue that no warming had occurred meant they had to discredit the 17,000 thermometer measurements, which looked to be the more reliable gauge of change.

In 2004, after years of complex research, climate scientists identified the source of the error, and it lay with the satellite data.[9] We learned earlier that the depletion of ozone was cooling the stratosphere, while greenhouse gases had been warming the troposphere. The satellites, it transpires, had been measuring both an increasingly warm troposphere and an increasingly cool stratosphere, and the meaningless average of these diverging temperatures was what was leading the researchers astray.[101]

While on the topic of the sceptics, it's worth considering one of the great, early enigmas in temperature trends. Between the 1940s and 1970s, despite increasing greenhouse gas levels in the atmosphere, Earth's average surface temperature declined. Furthermore, early global circulation models predicted that, with the amount of CO_2 released into the atmosphere over the century, Earth should have been warming twice as much as it actually did. Sceptics jumped on these anomalies both to discredit the computer models and to trumpet the idea that CO_2 and other greenhouse gases had nothing to do with rising temperatures. Both discrepancies, it turned out, resulted from a previously overlooked factor—the very powerful influence on climate of minuscule particles that drift in the atmosphere.

Known as aerosols, they can be anything from dust ejected by volcanoes to the cocktail of deadly particles originating from the smokestacks of coal-fired power stations. Desert landscapes produce them in large quantities, and diesel engines, tyre rubber and fires are all important sources. Early computer models did not include aerosols in their calculations, in part because no one fully appreciated the extent to which human activities were increasing their number. We now know that between one quarter and one half of all the aerosols in our atmosphere today are put there by human activity.[70]

Aerosols can be very damaging to human health. They were the cause of significant mortality in seventeenth-century London, and today, even with further developments, aerosols generated by burning coal kill around 60,000 people annually in the US.[81] Part of the reason is that coal acts like a sponge, soaking up mercury, uranium and other harmful minerals which are released when it is burned. The state of South Australia is home to the world's largest uranium mine, yet its largest single point source of radiation is not the mine but a coal-fired power plant at Port Augusta. It's no real surprise that lung cancers commonly result from burning coal. In Australia's Hunter Valley, where coal-fired power generation is concentrated, lung cancer rates are a third higher than in nearby Sydney, despite the pollution levels in the metropolis.

As a child I remember seeing No Spitting signs on the railway tunnel walls of my home city of Melbourne, and hearing stories of spittoons being used in my grandfather's day. When I travelled to China as an adult, and saw the inhabitants of grossly polluted cities such as Hefei hacking up foul congestion from their lungs, I realised that my forebears did not necessarily have worse hygiene habits than my generation. They simply battled with a cesspit-like atmosphere created by burning coal.

Scientists now think that the temperature decline of the 1940s to 1970s was caused by aerosols, with sulphur dioxide being particularly responsible. Sulphur dioxide is released when low quality coal is burned, and by the 1960s lakes and forests at high latitude in the Northern Hemisphere were dying. The trees were losing their needles, while the lakes were becoming crystal clear and emptied of life. The cause was acid rain resulting from the sulphur dioxide emissions of coal-burning power stations. This realisation prompted legislation to enforce the use of 'scrubbers' on coal-burning power plants in the industrialised world. These have been used since the 1970s and have dramatically reduced sulphur dioxide emissions.

There was, however, an unintended consequence of this. Aerosols of sulphate are most effective at reflecting sunlight back into space, and thus act powerfully to cool the planet. Because most aerosols last just a few weeks in the atmosphere (with sulphur dioxide degrading at the rate of 1–2 per cent per hour at normal humidity), the effect of installing scrubbers was immediate. As the air cleared, global temperatures, driven by CO_2 released from those very same power stations, resumed their upward creep.[58] The experience was the perfect example of how, in our Gaian world, everything is connected to and influences everything else.

The 1991 eruption of Mt Pinatubo in the Philippines provided an exceptional test of the new global circulation models' capacity to predict the influence of aerosols. It ejected 20 million tonnes of sulphur dioxide into the atmosphere, and a group led by NASA scientist James Hansen forecast that the result would be around 0.3°C of global cooling—and this figure is *exactly* what was seen in the real world.

Among the most important and best supported of these models' predictions are that the Poles will warm more rapidly than the rest of the Earth; temperatures over the land will rise more rapidly than the global average; there will be more rain; and extreme weather events will increase in both frequency and intensity. Changes will also be evident in the rhythms of the day and, as first predicted by Arrhenius, nights will be warmer relative to days, for night is when Earth loses heat through the atmosphere to space. There will also be a trend towards the development of semi-permanent El Niño-like conditions which, as we have seen, will have broad impacts.

We must now turn to the key uncertainty that remains in all models: will a doubling of CO_2 lead to a 2°C or 5°C increase in warming, and can we expect a reduction in this uncertainty in the near future? This is a critical issue not least because the current US government has signalled that it will not reconsider its climate change policy until there is more certainty. Given that almost thirty years of hard work and astonishing technological advances have failed to reduce the degree of uncertainty, we should not be too sanguine about hopes for more precision. Many would argue that we already know enough: even 2°C of warming would be catastrophic for large segments of humanity.

The most recent study of climate change, and the largest ever under-taken, was published in early 2005 by a team led from Oxford University. It was conducted by using the downtime on more than 90,000 personal computers, and it focused on the temperature implications of doubling CO_2 in the atmosphere. The average result of the many runs made indicated that this would lead to 3.4°C of warming. Overall, however, there was an astonishingly wide range of possibilities—from between 1.9 and 11.2°C of warming, the higher end of which had not been predicted earlier.[213]

As I read these results, an anomaly that had long niggled at me resur-faced. At the end of the last ice age CO_2 levels increased by 100 parts per million, and Earth's average surface temperature rose by 5°C. This suggests that CO_2 is a powerful influence on global temperature. Yet in most computer analyses, an increase in CO_2 almost three times as large (doubling pre-industrial levels) is predicted to result in a temperature rise of only 3°C.

This anomaly has serious implications for the survival of our civilisa-tion and of countless species. Scientists now working on aerosols think that they might have the answer. Direct measurement of the strength of sunlight at ground level, and worldwide records of evaporation rates (which are influenced primarily by sunlight) indicate that the amount of sunlight reaching the Earth's surface has declined significantly (up to 22 per cent in some areas) over the last three decades. It is as if we had been stopping-up that small 'window' in the atmosphere through which visible light passes.

This phenomenon is global dimming, and there are two ways that it operates: aerosols such as soot increase the reflectivity of clouds, and the contrails left by jet aircraft create a persistent cloud cover. Soot particles change the reflective properties of clouds by fostering the formation of many tiny water droplets rather than fewer, larger ones; and these tiny water droplets allow clouds to reflect far more sunlight back into space than do larger drops. The story with contrails is different. In 2001, in the three days following September 11, the entire US jet fleet was grounded, over which time climatologists noted an unprecedented increase in daytime temperatures relative to night-time temperatures. This resulted, they presume, from the additional sunlight reaching the ground in the absence of contrails.

If 100 parts per million of CO_2 really can raise surface temperature by 5°C, and if aerosols and contrails have counterbalanced this so that we have experienced only 0.63°C of warming, then their influence on climate must be enormously powerful. It is as if two great forces—both unleashed from the world's smokestacks—are tugging the climate in opposite directions, only CO_2 is slightly more powerful.

This leaves us with a grave problem, for particle pollution lasts only days or weeks, while CO_2 is difficult to clean up and lasts a century or more. So what does a 2°C or a 5°C rise in temperature mean—on the ground—to various peoples and ecosystems? These are questions to which we will return, but for now, this much can be said: if our understanding of global dimming is correct, then we only have one option. We *must* start extracting CO_2 from the atmosphere.

Before going any further, we must learn what questions can and can't be answered by computer models. One of the most fundamental human responses to any change is to ask what caused it. Earth's climate system, however, is so riddled with positive feedback loops that our usual concepts of cause and effect no longer hold. Consider that oft-used example from chaos theory, of the flutter of a butterfly's wing in the Amazon causing a cyclone in the Caribbean. But saying that something has simply caused

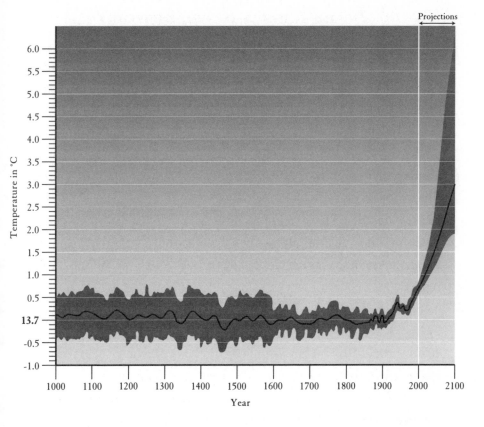

This graph, known as the 'hockey stick', shows trends in the average surface temperature of Earth from AD 1000 to 2100. Prior to 1900 this was 13.7 degrees. The grey area conveys uncertainty, which is reduced around 1850 when the thermometer grid was established. The projections on the right give a range of probable temperature increases to 2100.

something else is an unhelpful way of thinking. Instead what we have are seemingly insignificant initial occurrences—such as an increase of atmospheric CO_2—that lead to runaway change.

Another natural response is to ask what all of this means to me, and the area I live in, in the near future. Because weather conditions vary greatly day to day and year to year, there is no foolproof way of determining this.[23] From the perspective of a human lifetime, global warming is slow—making each decade a little warmer than the one before it— while climatic variability year to year, or even weather variability day to day, may be far greater than the climate change engendered by a shift in the decade-long average. In this regard, predicting the weather is very different from predicting the impacts of climate change: weather is best predicted for a specific location over a very short period—say a day or three. Climate change, in contrast, is best done on the global scale and many decades in advance.

A number of climate groups—often at the request of governments needing advice on how to prepare themselves—have produced computer-based projections for various regions of Earth and for time scales as short as a few decades. Three examples of such studies help provide a taste of the numerous regional predictions that abound. It's worth keeping in mind, however, that many climatologists question the usefulness of such reports.

Some of the most sophisticated regional predictions are those produced by the Hadley Centre for the climate of the United Kingdom from the 2050s to the 2080s.[104] They assume the presence of a range of greenhouse gas emissions from low to high. Low means strenuous measures to reduce CO_2 emissions would have been successfully implemented, and luck with feedback loops; high means business as usual and bad luck with feedback loops.

With all scenarios they discovered that by 2050 human influences on the climate will have surpassed all natural influences. In other words, there will be no more climatic 'acts of God', only human-made climate disasters. They predict that snow cover will decrease by up to 80 per cent near the British coast, and up to 60 per cent in the Scottish highlands. Winter rainfall is predicted to increase by up to 35 per cent, with more intense rainfall events, while summer rainfall will decrease, and one summer out of three will be 'very dry'. An event akin to the severe summer of 1995 (which had seventeen days over 25°C and four days over 30°C) may recur twice per decade, while the great majority of years will be warmer than the record-breaking 1999.[104]

The changes felt over the European landmass will be more extreme than the increase in the global average. Indeed, a global rise in surface temperature of just 2°C would bring a temperature rise to all of Europe,

Asia and the Americas of 4.5 degrees.[103] For Britain, this means a more Mediterranean-like climate and, as some newspapers put it, 'the end of the English garden'. More important are the challenges it throws up for matters such as water security, flood mitigation and human health.

In 2003 and 2004 two further regional studies, prepared by scientists from Stanford and the University of California at Los Angeles, focused on climate impacts for California.[163] They postulated that global warming would bring much hotter summers to the state and a depleted snowpack, threatening both water supplies and health. By the end of the century, heatwaves in Los Angeles would be two to seven times as deadly as today, snowpacks would decline by half or more, and three-quarters to nine-tenths of all of California's alpine forests would be lost.

The third example focuses on the state of New South Wales, with predictions made by Australia's leading science research body, the CSIRO. The time scale used here is short—only three decades in some instances— and it utilises twelve separate climate simulations that deliver a wide range of possibilities. These include temperature increases across the state of between 0.2 and 2.1°C while the number of cold spells, and thus frosts, will reduce. The number of very hot days (above 40°C) will increase, as may winter and spring droughts, extreme rainfall events and wind speeds, and there will be changes in wave patterns and possibly the frequency of storm tides.[207]

In reading regional reports it is evident that the shorter the time scale the less certain are the predictions; conversely, the longer the time scale and the larger the region involved, the more they resemble global models, which are our best source of information. There is one other very important reason why short-term studies are not so significant. The gas is already in the air and right now we have no way of getting it out. This means that the course of climate change is set for at least the next several decades.

SEVENTEEN

THE COMMITMENT, AND APPROACHING EXTREME DANGER

What makes global warming so serious and so urgent is that the great Earth system, Gaia, is trapped in a vicious circle of positive feedback. Extra heat, from any source, whether from greenhouse gases, the disappearance of the Arctic ice or the Amazon forest, is amplified, and its effects are more than additive. It is almost as if we had lit a fire to keep warm, and failed to notice, as we piled on the fuel, that the fire was out of control and the furniture had ignited. When that happens, little time is left to put out the fire. Global warming, like a fire, is accelerating and almost no time is left to act.

James Lovelock, *Independent*, 24 May 2004.

Researchers at the Hadley Centre talk of a 'physical commitment to climate change'.[164] This refers to the fact that the full impact of the greenhouse gases already in the atmosphere will not be felt until around 2050, the implication being that, if greenhouse gas emissions stopped immediately, Earth would reach a new stable state, with a new climate, around 2050. Because we lack a way to take greenhouse gases out of the atmosphere, this five-decade period of 'catch up' is a true physical commitment due to the long life of CO_2 in the atmosphere. Much of the CO_2 released when our great-grandmothers stoked their coal-fuelled stoves in the aftermath of World War I is still warming our planet today. Most of the damage was done, however, starting from the 1950s, when our parents and grandparents drove about in their fin-tailed Chevrolets and powered their labour-saving household appliances from inefficient coal-burning power stations. But it is the baby-boomer generation that is most culpable: half of the energy generated since the Industrial Revolution has been consumed in just the last twenty years.

It's easy to condemn the extravagance that led to the situation in which we now find ourselves, but we must remember that until recently nobody had the slightest idea that their tailpipe emissions or Hoover vacuum cleaner would have an impact on their children and grandchildren. The same cannot be said for us today, for the true cost of our four-wheel-drives, air conditioners, electric hot water service, clothes dryers and refrigerators is increasingly evident to all. Moreover, in many developed nations we are three times as affluent on average as our parents were at the same stage of life, and therefore we are able to bear the cost of changing our ways.

A closer look at inertia in the Earth's climate systems is necessary to understand more fully what our 'commitment' actually means. As we found earlier, the atmosphere, land surface and oceans all respond at different rates to increases in greenhouse gases. In 2002 the surface temperature of the planet as a whole was 0.8°C above pre-industrial levels, the land surface was 1.2°C warmer, while the troposphere at one to eight kilometres above our head (as measured by satellites) was 0.25°C warmer than it was on average over the preceding twenty years; different parts of the Earth system vary in their response to warming, and distributing the extra heat is one reason for the lag.

Our commitment is also influenced by the CO_2 we have already released, the positive feedback loops that amplify climate change, global dimming and the speed at which human economies can decarbonise themselves. Of these, the first—existing greenhouse gas volumes—is known and gives us our 'existing commitment'. The second and third—positive feedback loops and global dimming—are still being explored by scientists. And the fourth—the rate at which we humans can change our emissions—is being argued over right now in parliaments and boardrooms around the world. It is also the only impact over which we have control.

Scientists say that a 70 per cent reduction in CO_2 emissions from 1990 levels by the middle of the twenty-first century is required to stabilise Earth's climate. This would result in an atmosphere with 450 parts per million CO_2, and our global climate stabilising by around 2100 at a temperature at least 1.1°C higher than the present, with some regions warming by as much as 5°C. The European nations are talking of emissions cuts on this scale, but given the intransigence of the coal industry and the policies of the current US administration this may be unachievable as a global target. A more realistic scenario may be stabilisation of atmospheric CO_2 at 550 parts per million—double the pre-industrial level. This would result in climatic stabilisation centuries from now, and an increase in global temperature of around 3°C this century, give or take a couple of degrees (the 'give' more likely than the 'take'). But remember that even this depends on good luck, for despite our best efforts the level of greenhouse gases already in the atmosphere may trigger positive feedback loops with the potential to destabilise the carbon cycle.

So how does this commitment relate to concerns that Earth may cross some threshold of climate change, beyond which lies extreme danger? The United Nations Framework Convention on Climate Change states its ultimate objective as the stabilisation of greenhouse gases at a level that 'would prevent dangerous anthropogenic interference with the climate system'. This means that climate change should not proceed at a rate faster than that to which ecosystems and food production systems can adapt, and also at a rate that does not threaten economic development.[188] But what is that rate? What is the threshold of dangerous 'climate change'?

In 2002 University of Maryland economist Thomas Schelling, who is an apologist for the US's refusal to ratify Kyoto, put it at 'probably between 600 and 1200 parts per million'.[123] This means somewhere in the range of 2 to 9°C of surface temperature increase. A more widely accepted view places the limit at around 2°C of warming. As 0.63°C of warming has already occurred, this leaves us a latitude of around 1.3 degrees of temperature increase. Yet Michael Mastrandrea and climatologist Steven Schneider write:

> It is possible that some thresholds for dangerous anthropogenic interference with the climate system are already exceeded, and it is likely that more such thresholds are approaching…despite the great uncertainty in many aspects of integrated assessment, prudent actions can substantially reduce the likelihood and thus the risks of dangerous anthropogenic interference.[56]

In other words, it's too late to avoid changing our world, but we still have time, if good policy is implemented, to avoid disaster. Good policy, in Mastrandrea and Schneider's model, means a carbon tax of US $200 per tonne, implemented by 2050, which is sufficient to reduce the probability of dangerous climate change to zero.[56]

Perhaps a more useful way of looking at the problem is to quantify *rates* of change that are dangerous. After all, life is flexible, and if given sufficient time it can adapt to the most extreme conditions. So it is the rate, not the direction or overall scale of change, that is important. Climate scientists taking this line argue that 'warming rates above 0.1°C per decade are likely to rapidly increase the risk of significant ecosystem damage'.[188] Similarly, rates of sea level rise above two centimetres per decade would be dangerous, as would a rise of five centimetres overall.[188] But the question of what constitutes dangerous climate change raises another question—dangerous to whom? For the Inuit, whose primary food sources of caribou and seal are now difficult to find as a result of climate change, an economically and culturally damaging threshold has already been crossed.

When we consider the fate of the planet as a whole, we must be under no illusions as to what is at stake. Earth's average temperature is around 15°C, and whether we allow it to rise by a single degree, or 3°C, will decide the fate of hundreds of thousands of species, and most probably billions of people. Never in the history of humanity has there been a cost-benefit analysis that demands greater scrutiny.

EIGHTEEN

LEVELLING THE MOUNTAINS

> Oh God, that one might read the book of fate,
> And see the revolution of the times
> Make mountains level, and the continent,—
> Weary of solid firmness,—melt itself
> Into the sea!
>
> William Shakespeare, *Henry VI Part 2*

For every hundred metres you climb a mountain, the temperature drops by more than half a degree Celsius. Without this cooling, mountains would be nothing more than topographically challenged versions of the lowlands that surround them. It's in this sense—by threatening to make them biological clones of the surrounding lowland regions—that climate change can level the world's mountains. That process can be seen most clearly now in the fate of the world's tropical glaciers and snow-dusted peaks. These habitats are already restricted to the summits, and we are 'committed' to losing them, for neither the snows of Kilimanjaro nor the glaciers of New Guinea can survive current levels of CO_2 for more than a couple of decades. And below those icy realms, every habitat—from alpine herbfields to elfin woodlands and mossy mid-montane forests—each of which has its own unique species, is climbing upwards.

Nothing in predictive climate science is more certain than the extinction of many of the world's mountain-dwelling species. We can even foretell which will be the first to go. This high degree of scientific certainty comes from three factors. First, the effect of rising temperatures on mountain habitats is easily calculated, and past adjustments in response to warming are well documented. Second, the conditions that many mountain-dwelling species can tolerate are known. And finally, as the climate warms mountain species have nowhere to go but up, and the height of mountain peaks worldwide has been precisely ascertained. Given the rate of warming, we can calculate the time to extinction of most mountain-dwelling species.

The last time the world warmed rapidly—at the end of the last ice age—the retreat of species to higher, cooler regions was swift and inexorable. On the island of New Guinea alpine herbfields, which are now mostly restricted to elevations above 3900 metres (the tree-line), then occurred as low as 2100 metres. This mile-high retreat reduced their aerial extent by nine-tenths, and today they can only be found on the summits of the island's highest mountains—secluded jewels in an otherwise forested landscape. The reason for their ascent was an increase in global surface temperature of around 5°C over 7000 years.

We know that our planet must heat by 1.1 degrees this century come what may, and business as usual will commit us to a 3°C increase in temperature. The very highest peak in New Guinea—Puncak Jaya—is just under 5000 metres, which means that, taking past changes as a guide, even a rise of 3°C would push the last of New Guinea's alpine habitat off its summit. Indeed, given such extreme changes, there are few mountains anywhere on Earth high enough to provide an alpine refuge.

Waking in the crisp air on a New Guinea mountaintop and seeing delicate spiderwebs strung between tree-ferns, glittering with dew, is an experience to cherish. In the slanting light the dominant colours of these open, equatorial meadows are bronze and brilliant green, interspersed with the flamboyant red, orange and white flowers of tussock-like rhododendrons and unique orchids. Around your feet in the mossy soil are the scratchings of the metre-long long-beaked echidna (*Zaglossus bartoni*)—the largest egg-laying mammal on Earth—and burrows of the alpine woolly rat (*Mallomys gunung*), which at nearly a metre long from nose to tail-tip is also a giant.

At dawn the air is full of birdsong, for these mountains are the retreat of birds of paradise, parrots, and hordes of honeyeaters that flock to the flower-filled bushes. Towards mid-morning, from the scattered marshes you'll hear *Oooh, ooh*, which you might think (as I did) sounds like your favourite maiden aunt, tipsy after Christmas dinner. But here is a tiny rose-pink frog—no bigger than a child's thumbnail—and so new to science that it hasn't yet been named.

Every high tropical mountain on Earth has an equivalent alpine habitat that is rich in biodiversity, and below them are mountain forests that are even richer in life. Indeed, the world's mountain ranges nurture a staggering variety of life—from iconic species such as pandas and mountain gorillas to humble lichens and insects. On the global scale, the importance of mountain habitats is illustrated by diversity in the alpine zone—the area lying between the tree-line and the perpetual snow of the mountaintops. This region of shrubs, tussock grasses and herbfields usually has a highly endemic fauna and flora. Although alpine habitats make up a mere 3 per cent of the surface of the Earth, they are home to over 10,000 plant species, along with countless insects and larger animals; they are mega-diverse regions.

The study that identified the global 'fingerprint' of climate change found that, over the course of the twentieth century, mountain-dwelling species have withdrawn on average 6.1 metres up the slopes of their montane homes per decade.[27] The creatures and plants did this because conditions at the lower margins of their distributions became intolerable—too hot or dry—or because of newly arrived species with which they could not compete. This may seem a small amount of movement, but we must remember that our planet has not been warmer than it is now for millions of years, a situation that has left many ancient species clinging to the last few hundred metres of mountain peaks around the world.

Few studies of the impacts of climate change on specific mountain regions have been completed, perhaps because the work is too depressing. The most detailed thus far are those undertaken by Steve Williams of James Cook University and his colleagues, which deal with the impact of climate change on the rainforest-clad mountains of northeast Queensland.[35]

These mountain ranges are centred on the Atherton Tablelands west of Cairns, and cover 10,000 square kilometres. Yet despite their small size they are arguably the most important habitat in all of Australia, for they are home to an archaic assemblage of plants and animals—survivors from the cooler, moister Australia of 20 million years ago. The significance of this region to the world as a whole was recognised in 1988 when the rainforests were listed as Australia's first World Heritage Area. Tourists now flock to the area and one of the most popular activities is a night walk, when an abundance of marsupials can be seen at close range by spotlight. In some places the forest is alive with grunts, squeals and rustlings.

High up in the tallest trees you'll hear lemuroid ringtail possums leaping from branch to branch. They are living fossils—remnants of the lineage that gave rise to the majestic metre-long greater glider of the eucalypt forests. Lemuroids lack a gliding membrane, but are extraordinary leapers whose noisy crashing through the canopy is one of the most constant noises at night. Lower down in the trees you might see the green ringtail possum with her large young. So selective are they in their diet that to learn which leaves are best the young stays with its mother until it's almost adult-sized. The reason that such creatures haunt the mountain summits is clear. Just four to five hours spent in temperatures of 30°C or above will kill them, and such temperatures are an almost daily event in the surrounding lowlands.

Sixty-five species of birds, mammals, frogs and reptiles are unique to the region and none can tolerate warmer conditions. They include the golden bowerbird (*Prionodura newtoniana*), the Bloomfield nursery frog (*Cophixalus exiguus*) and Lumholtz's tree-kangaroo (*Dendrolagus lumholtzi*). It's not widely known outside of Australia that some kangaroos inhabit the treetops of tropical rainforests, but such creatures were once widespread, for their fossils have been found as far south as Victoria. Today they survive only in the rainforests of northeast Queensland. They and the other rainforest species owe their prehistoric decline to a conspiracy of tectonic and climatic forces. Forty million years ago continental drift began to move Australia northward, and the additional warmth and a changing climate dried out the continent, banishing the cool rainforests to the east coast. Then the ice ages devastated the southern forests, leaving only the rainforests of northeast Queensland as a refuge.

Steve Williams's study indicates that rising temperatures will directly affect creatures such as the green ringtail which need to stay cool, and that periods of extreme temperature will become more common. Then there is the impact that higher CO_2 levels will have on plant growth. Plants grown experimentally in CO_2-enriched environments tend to have reduced nutritional value, tougher leaves and higher concentrations of defensive chemicals (such as tannins and phenolics), making them a much poorer food source. This change alone is predicted to reduce possum density, and as species become restricted to ever-higher elevations the very poor soils that dominate on the summits will reduce the nutritional value of their food further. If this were not enough, rainfall variability is likely to increase, with droughts becoming more pronounced, while the cloud layer, which now provides 40 per cent of the water that nourishes the mountain forests, will rise, exposing the forests to more sunlight and thus evaporation. All of this adds up to a catastrophic impact.

With a rise in temperature of just 1°C (which will occur no matter what we do) at least one unique wet tropics species—the Thornton Peak nursery frog (*Cophixalus* sp.)—will become extinct. This is a tragedy, for so recently was this creature discovered that it hasn't yet received a scientific name. With a 2°C increase the wet tropics ecosystems will begin to unravel. At a 3.5°C increase, around half of the sixty-five species of animals unique to the wet tropics will have vanished, while the rest will become restricted to tenuous habitats of less than 10 per cent of their original distribution. In effect their populations will be non-viable, their extinction only a matter of time.

The implications of Williams's studies for the future of Australia's biodiversity are outrageous. The sixty-five species of larger creatures unique to the wet tropics constitute just the tip of a mountain full of biodiversity. Consider the native pines, which are a tiny proportion of the region's flora. Two species with fern-like foliage and glorious red or blue fleshy fruit (*Prumnopitys amara* and *P. ladei*) are restricted to the summits of the ranges, while the bunya pine (*Araucaria bidwilli*)—a relative of the monkey puzzle tree and the oldest species in an ancient lineage—is restricted to two mountain ranges. This species, or something like it, has been around since the Jurassic Era some 230 million years ago. Its loss would be calamitous; yet in many instances we would have no idea of what we are losing, for in 1994 an entirely new genus of rainforest tree was found on the ranges' highest peaks—Mt Bartle Frere and Mt Pieter Botte.[208] A distant relative of the banksias and proteas, it has a hard, nut-like fruit, fossils of which occur in 30-million-year-old deposits in Victoria. These examples are in no way complete when one remembers the diversity in orchids, ferns and lichens. And I haven't touched on the invertebrates yet—those legions of worms, beetles and other flying and crawling things found in their tens of thousands.

The impending destruction of Australia's wet tropics rainforests is a biological disaster on the horizon, and the generation held responsible will be cursed by those who come after. What will they tell their children if their increasingly large homes, four-wheel-drives and refusal to ratify Kyoto cost them the nation's foremost natural jewels?

Throughout the world, every continent as well as many islands has mountain ranges that are the last refuge of species of remarkable beauty and diversity. And we stand to lose it all, from gorillas to pandas to New Zealand's vegetable sheep (a unique tussock plant). No rescue effort could ever be comprehensive enough to establish captive colonies of even one tenth of 1 per cent of the species under threat. There is only one way to save them. We must act to stop the problem at its cause—the emission of CO_2 and other greenhouse gases.

There is, surprisingly, one group of species that will benefit enormously from this aspect of climate change. These are the parasites that cause the four strains of malaria. As rainfall increases the mosquitoes that carry the parasite will spread, the malarial season will lengthen and the disease will proliferate. From Mexico City to Papua New Guinea's Mt Hagen, the mountain valleys of the world support human populations in high densities. And they are healthy, glorious places in which disease, where population density is not too great, is rare. Just below these communities—in the case of New Guinea at around 1400 metres—are great forests where no one lives. This is because of malaria, which is so prevalent in parts of the tropics that it controls human populations. In the near future, global warming will grant access to the malarial parasite and its vector the *Anopheles* mosquito to those high mountain valleys, and there they will find tens of thousands of people without any resistance to the disease.[196]

NINETEEN

HOW CAN THEY KEEP ON MOVING?

> They looking back, all th' eastern side beheld
> Of Paradise, so late their happy seat...
> The world was all before them, where to choose
> Their place of rest...
> John Milton, *Paradise Regained,* Book XII

Researchers Camille Parmesan and Gary Yohe defined the 'global finger-print' of climate change. But what will that fingerprint look like after 1.1°C or even the predicted average 3°C of warming?

Species have survived past shifts of climate change because mountains have been tall enough, continents extensive enough and the change gradual enough for them to migrate. Sometimes the distances travelled have been enormous. Just 14,000 years ago, for example, the deciduous forests that now grow around Montreal in Canada could be seen only in northern Florida. The climatic changes that prompted that migration, although far slower, were on a scale similar to those projected to occur this century. This tells us that the key to survival in the twenty-first century will be to keep on moving. But how will species manage such long-distance movements in the modern world?

The problem presented by this aspect of climate change was, for plants at least, first outlined in 1996 by a group of Australian botanists led by Lesley Hughes of Macquarie University.[159] As early as 1992 it was realised that, as a consequence of climate change, temperatures in Australia may rise as much as 5°C in response to a global increase of just 2°C.

Concerned by the impact on Australia's biodiversity, Hughes examined the distribution of 819 species of *Eucalyptus*, and found that, although these trees characterise the Australian landscape, the majority of species occupy small discrete regions defined by very narrow temperature zones. More than 200 species (25 per cent) have ranges spanning just 1°C, while 41 per cent span just 2°C. In fact, in 75 per cent of them the range is less than 5°C. Should Australia's temperature rise by only 3°C over this century (which is a realistic figure if we go on with business as usual) half of Australia's *Eucalyptus* species would grow outside their current temperature zone.[159] If they are to survive they must migrate, yet numerous barriers, including the Southern Ocean, and human modified landscapes stand in the way.

In 2004 news was broadcast that the eucalypt forests of Tasmania's World Heritage areas were dying as a result of drier, hotter conditions. To see Dr Hughes's prediction of the fate of the gum tree being fulfilled less than a decade after it was made, and in the very region of Australia where (because of its high latitude) climate change is progressing most rapidly, was frightening.

William Hare, on behalf of the Potsdam Institute, has written a global view of the impacts likely to be seen in the world's natural systems as a result of shifting climate.[188] Looking at the tabulated results, it's clear that there is not an ecosystem on Earth that will be unaffected by climate change. Some environments, however, are threatened by even small shifts.

South Africa's succulent Karoo flora comprises some 2500 species of plants found nowhere else—the richest arid-zone flora on Earth—and it's renowned for the beauty of its spring flowers, which depend on marginal winter rainfall. As the climate changes, there's simply nowhere for this vegetation to migrate to, for to the south and east—the direction which climate change will drive it—lie the Cape Fold Mountains, whose topography and soils are unsuitable for Karoo plants. Computer simulations indicate that by 2050, 99 per cent of the succulent Karoo will have vanished.

To the south of the Cape Fold Mountains is the fabulous fynbos, one of six floral kingdoms on Earth, and the most diverse plant community to be found outside the rainforests. The plants are little more than knee-high, but their form is extraordinary. Rushes bear brilliant bell-shaped flowers, whose nectar is sipped by brightly-coloured 'humming flies' with two-centimetre-long siphons which reach deep into the bells. Rocky slopes are adorned with bushy king proteas studded with saucer-sized, pink star-flowers, while the profusion of pea flowers, daisy-like forms and iris relatives seems endless. Hemmed in by ocean at the southern tip of the continent, the fynbos is a natural paradise. But as the Earth warms, its azure backdrop means that it has nowhere to go, and so it will lose over half of its extent by 2050, and along with that a significant number of its 8000-odd endemic species.

The diverse heathlands of Australia's southwest comprise over 4000 species of flowering plants. With just half a degree of additional warming, the fifteen species of mammals and frogs that have been studied and are unique to the region would be restricted to tiny relictual habitats, or would become extinct. Few of the plant groups have been studied in detail, though one exception is the genus *Dryandra*. Two-thirds of the ninety-two species of these banksia-like shrubs and small trees would become extinct with such a shift. Yet we already know that half a degree of warming is inevitable.[28]

It's the topography of this region and its history of land-clearing that make it so vulnerable. Climate change will drive the diverse plant communities ever southwest to the ocean. Yet those that can move at all will be the lucky ones, for much of the southwest today is a vast wheat-field. Some species survive only on roadside verges, along railway lines, or in pocket-handkerchief-sized flora reserves. A few exceptional areas have been set aside as larger national parks, but in the face of galloping climate change even these will become little more than death traps.

The critical point being made here is that global warming could not have come at a worse time for biodiversity. In the past when abrupt shifts of climate occurred, trees, birds, insects—indeed entire biotas—would migrate the length of continents as they tracked conditions suitable for them. In the modern world, with its 6.3 billion humans, such movements are not possible. Today, most biodiversity is restricted to national parks and forests that are often surrounded by an immensity of landscape profoundly modified by human activity.

While the Mediterranean-type plant communities of South Africa and Australia are especially vulnerable in the face of climate change, enormous changes will occur almost everywhere.

Because of drying trends over the American west, rising seas and increased storm surge, the wintering habitat for migratory shore birds in North America will be significantly reduced. Warmer summers, higher evaporation rates and more variable weather dictate that the breeding habitat of waterfowl in regions such as the Prairie Potholes will suffer. Warming streams mean that salmon will diminish in number, while in the North Atlantic commercially valuable fish are already following the cold water downwards and northwards. Mexico's fauna will be squeezed by heat, drying and extreme weather events, resulting in many extinctions, and these same factors have led botanists to declare that one third of Europe's plant species face severe risks.

On smaller landmasses the situation is even more dire. Because waves of climate change will wash over islands, making them unsuitable habitat for many residents, many Pacific island birds will be pushed beyond their limits, and there will be extinctions in all forms of life from island trees to unique insects. And as we have seen, national parks are now islands in a sea of human modified environments. Kruger National Park in South Africa is nearly the size of Israel, yet it still stands to lose two-thirds of its species.[188]

These, it must be remembered, are just a few examples of projected biodiversity loss from regions that have been studied. Imagine the world's climate zones changing dramatically over your lifetime—so that Washington's climate is more akin to that of Miami today—and try to think what it means for the forests, birds and other animals of the region where you live, and you'll begin to see the bigger picture.

I was in London when I undertook some of this research, and one jet-lagged morning I awoke before the dawn and sat, watching the lightening of the eastern sky. Slowly a familiar shape took form. It was a gum tree, growing sturdily in a region traditionally far too cold for it to thrive. Then, as the first fingers of dawn probed the small garden below, a group of birds alighted in the tree. They were Indian parakeets. I had been expecting to see sparrows, but was told that they were all but extinct in the city. It made me wonder what the climate-shocked city of the future might be like.

There is another way to try to understand how climate change is affecting the planet's ecosystems. We can mass together the available data, which involves observations of over 1000 species such as trees, crustaceans and mammals, to see what it says statistically, as a whole. This was the approach taken by a group of researchers led by Chris Thomas of the University of Leeds, who published their findings in *Nature* in late 2004.

The project examined the fate of 1103 species of plants and animals, from proteas to primates, in the face of climate change to 2050. The locations were drawn from regions covering 20 per cent of the Earth's surface, including Mexico, South Africa, Europe, South America and Australia.

Thomas and his colleagues found that, at the lowest degree (inevitable) of global warming—between 0.8 and 1.7°C—around 18 per cent of the species sampled will, in the dispassionate language favoured by science journals, be 'committed to extinction'—in other words doomed. At the mid-range predictions—1.8–2.0°C—around a quarter of all species will be extirpated, while at the high range of predicted temperature rises (over 2°C) over a third of species will become extinct.

Believe it or not this is the good news; in these analyses it is assumed that the species can migrate. But what chance does a protea have of dispersing across the populated coastal plain of South Africa's Cape Province, or a lion-tamarin monkey crossing the agricultural fields that have all but obliterated the Brazilian Atlantic rainforests? The answer is very little indeed, and for species that cannot disperse the likelihood of extinction is roughly doubled. This means that at the high range of predicted temperatures, over half (58 per cent) of the 1103 species examined are 'committed to extinction'.[33]

Extrapolating from Thomas's data set it appears that at least one out of every five living things on this planet is committed to extinction by the existing levels of greenhouse gases. The World Wildlife Fund, the Sir Peter Scott Trust and the Nature Conservancy have worked for decades to save, in real terms, relatively few species. Now it seems countless thousands will be swept away by a rising tide of climate change unless greenhouse gas emissions are reduced.

We must remember, however, that if we act now it lies within our power to save *two* species for every one that is currently doomed. If we carry on with business as usual, in all likelihood three out of every five species will not be with us at the dawn of the next century.

TWENTY

BOILING THE ABYSS

Let us think of them that sleep
Full many a fathom deep
        Thomas Campbell, 'The Battle of the Baltic'

When marine biologists trawl the ocean's depths and haul up the bizarre creatures that reside there, the animals—unavoidably—are already dying. Black, toothsome bodies of deep-sea anglers lie inert, their luminescence slowing to a flicker, while predators such as the stoplight loosejaw (*Malacosteus niger*) grow pale and vomit up their last meal, which is often a fish larger than itself. Within minutes movement ceases, and the eyes of a creature that has been plucked from its element glaze over.

It's the change in pressure that killed them, scientists used to say, for this creature's world is one where the force of the kilometres-high column of water overhead is so intense that a submarine would buckle in an instant. As proof of this idea experts pointed to those few deep-sea fishes that have swim-bladders. They reach the surface grossly distorted, their air-sacs so swollen with expanding gas that their bodies stretch to bursting. Despite such gruesome 'proof', we now know differently.

In your imagination, grit your teeth and pick up a hairy seadevil (*Caulophryne polynema*) that has just emerged from a depth of three kilometres. Then toss its black, sack-like and filament-covered body (trust me, it is surely the most grotesque of all fishes) into a bucket of icy sea water. Now step back. Within minutes vitality will return to its frame, its great fang-studded jaws will snap, and the filament-clotted 'fishing rod' that protrudes from between its eyes will flicker. The creature has recovered from the trauma of its ascent, demonstrating that a moment ago its life was being threatened not by pressure but by warmth; they are denizens of the deep ocean water, where temperatures hover near zero. Even water temperatures that would chill us to death in minutes are fatally warm to these fish.

The structure of the world's oceans is critical to our climate. There are three layers, separated by their temperature. The top 100 metres or so vary enormously in temperature; near the Poles it can be below zero, while at the equator it can exceed 30°C. Below this familiar, light-filled world, to the depth of around a kilometre, is a zone of temperature transition—as you descend, so does the mercury in the thermometer. At around a kilometre down we have reached the world's deep ocean water, and from bottom to top it's remarkably stable in temperature—varying between –0.5°C (it can be below freezing and not turn to ice because of the salt) and 4°C. Most of the water in this lightless realm is exported from Antarctica, where it has been chilled to near freezing point by submarine currents.

Let's briefly consider the Poles, where the icy water of the deep ocean comes to the surface. Richard Feely of the Pacific Marine Environmental Laboratory and his colleagues have investigated what might happen in these regions as more CO_2 is absorbed. The oceans become acid, and because the ocean's buffer, carbonate, is in limited supply it's liable to drop below the level at which it can be used by shell-forming animals. After that point is reached the carbonate is leached out of the creatures' shells and back into the oceans, making it impossible for them to maintain their protective covers.[73]

Animals such as pectens and oysters, which use aragonite (calcium carbonate of a different structure from most mollusc shells) are especially vulnerable, because aragonite's dissolution threshold (the point where it dissolves in salt water) is around a third lower than that of calcite, but eventually even crabs, prawns and worms will suffer.

This problem may not occur for several hundred years, but by the time we see the first signs it will be far too late to do anything about it. The place to look for the first shell-less oysters is in the sub-Arctic north Pacific, for there the saturation point for carbonate is lower (just two-thirds that of the tropical ocean) than anywhere else.

Its first influences will be felt in winter, when low temperatures and wind-driven mixing of the surface and lower layers create the right conditions. Thereafter, the malaise will creep towards the equator where, in time, all species that secrete shells will be affected.[73] Because of the ocean's inertia, by the time the first signs of this shift are felt, it will be far, far too late to reverse it. If you wish your great-great-grandchildren and those who come after to know the taste of oysters, we need to limit CO_2 emissions now.

While most people would mourn the loss of oysters, they might equally feel that the world is well rid of such creatures as the hairy seadevil and the stoplight loosejaw; but the deep oceans are one of the most wondrous and extensive realms on our planet. They are also a last frontier where it's still possible to be surprised by a five-metre-long shark that is not just a new species, but an entire family new to science.

Such was the case with the megamouth (*Megachasma pelagios*), the first known example of which got entangled with the sea anchor of a US naval vessel working in 4.5-kilometre-deep water off Hawaii in the 1970s. These great sharks are filter feeders that—as far as we know—spend their entire lives balanced on their tails, migrating vertically in the trackless ocean.

If such monsters have remained unknown, imagine how many smaller creatures there are waiting to be discovered. And life in the depths is so specialised that it is sure to enlighten us as to how creatures survive at the extreme limits of habitability.

The gulpers are eel-like creatures which seem to be all mouth, stomach and tail, the tip of which is splendidly illuminated. They wait in the depths, their neon-like tail-tip curled round so that it sits just before their gulping maw, and when something comes to investigate—they seize it. Having grasped their prey tail-first, they must swallow what is often a spiny fish backwards, which they achieve by sliding their bodies slowly over the food, like sliding a sock over your foot. The umbrella-mouth gulper (*Eurypharynx pelecanoides*) has the largest mouth relative to its size of any backboned creature on Earth, yet so low on calcium is it that, after mating, it reasorbs its jaws and teeth in order to furnish its fertilised eggs with enough calcium to form embryonic skeletons.

Even more bizarre are the anglerfishes. The illuminated netdevil (*Linophryne arborifera*) is the most phosphorescent of all fish, its great beard and 'fishing rod' resembling a fully lit-up Christmas tree. That's the female, for the male is a useless parasite of a thing. When the size of a mosquito fish, he found his partner in life and bit into her belly. Now he is nothing but a parasitic testicle that feeds from her blood, and is stimulated into releasing his sperm when required.

The ocean deep is not just another of life's zones, it is almost a parallel universe brimming with evolutionary possibilities. What possible threat, you might wonder, could human activity pose to such a world? While the threat is not immediate, lessons from the past indicate that even this vast realm may fall victim to climate change.

Fifty-five million years ago, when an eruption of methane warmed our planet, the ocean depths became nearly as warm as its surface, and life in the abyss was almost annihilated. We have no remains of deep-sea fishes surviving from this time (indeed we have almost no fossils of them at all) but the surviving evidence in the rocks speaks eloquently of the mass extinction of the smaller creatures that shared their habitat.

Much of the diversity of the modern ocean depths is likely to have evolved since the Earth cooled around 33 million years ago, and a rapidly refrigerating Antarctica began to export icy sea water around the world. Although scientists are detecting a warming of the oceans at depth, it will take hundreds of years and a century or more of business as usual to heat them. In that possible runaway greenhouse world of the future, however, netdevils and gulpers might find themselves writhing in the agonies of heat-induced stress, even in the recesses of their dark realm.

TWENTY-ONE

THE PACK OF JOKERS

> Arguably, the largest oceanic change ever measured in the era of modern instruments is in the declining salinity of the subpolar seas bordering the North Atlantic.
>
> Daniel Glick, *National Geographic*, 2004.

Thus far we have considered only what might happen if present trends continue. But the fossil record reveals that even when the causes of climate change are slow and steady, things do not always go smoothly on planet Earth. Instead Earth's systems sometimes snap and a new world order is suddenly created, to which the survivors must adapt or perish.

There are three main 'tipping points' that scientists are aware of for Earth's climate: a slowing or collapse of the Gulf Stream; the demise of the Amazon rainforests; and the release of gas hydrates from the sea floor. All three occur on occasion in the virtual worlds of the global circulation models, and there is some geological evidence for all having happened in Earth history. This constitutes strong evidence that such events are possible, and that, given the current rate and direction of change, one, two or perhaps all three may take place this century. So what leads to these sudden shifts, what are the warning signs, and how might they affect us?

SCENARIO 1: COLLAPSE OF THE GULF STREAM

The importance of the Gulf Stream to the Atlantic rim countries is enormous. In 2003 Andrew Marshall, the originator of the Star Wars defence system and a grey eminence in the Pentagon, commissioned Peter Schwartz (former head of scenario planning for Royal Dutch Shell) and Doug Randall of Emeryville (a company specialising in scenario planning) to write a report outlining the implications for US national security should the Gulf Stream collapse. The purpose of the report was, its authors said, 'to imagine the unthinkable'. In order to do so they 'created a climate change scenario that although not the most likely, is plausible, and would challenge United States national security in ways that should be considered immediately'.[53]

Their scenario involved a slowing of the Gulf Stream as a result of fresh water from melting ice accumulating in the north Atlantic. It assumes that a slow warming of the planet will continue for another six years (until 2010), but then a dramatic shift will occur—a 'magic gate' that will abruptly alter the world's climate.

As a result of this shift, their 'weather report' for 2010 predicts persistent drought over critical agricultural regions, and a plunge in average temperatures of more than 3°C for Europe, just under 3°C for North America, and 2°C increases for Australia, South America and southern Africa.

In imagining the human response to such rapid change, the report's authors draw on the work of Harvard archaeologist Steven LeBlanc, who describes the relationship between human 'carrying capacity' and warfare thus: 'Humans fight when they outstrip the carrying capacity of their environment', and 'Every time there is a choice between starving and raiding, humans raid'.

Perhaps with their Pentagon readers in mind, Schwartz and Randall also anticipate nuclear weapons will proliferate and global co-operation will break down through the growing pressure for survival. Only the most combative societies, they say, will survive, and within these societies things look little better. Attitudes will change:

> As famine, disease, and weather-related disasters strike...many countries' needs will exceed their carrying capacity. This will create a sense of desperation...Perhaps the most frustrating challenge...is that we'll never know how...many more years— 10, 100, 1000—remain before some kind of return to warmer conditions.[53]

The impacts are compounded by the projected lack of co-operation between nations in the face of the disaster; mass starvation would be followed by mass emigration as regions as diverse as Scandinavia, Bangladesh and the Caribbean become incapable of supporting their populations. New political alliances would be forged as a scramble for resources ensues, and the potential for war is greatly heightened.

By 2010–20, with water supplies and energy reserves strained, Australia and the US would focus increasingly on border protection to keep out the migrating hordes from Asia and the Caribbean. The European Union, the report says, may go one of two ways—either becoming unified with a focus on border protection (to keep out those homeless Scandinavians, among others), or driven to collapse and chaos by internal squabbling. And they posit that Russia, made suddenly acceptable by its huge energy resources, may join the EU. The report makes seven recommendations to the US government to prepare for such eventualities, including exploring geo-engineering options (such as CO_2 sequestration) that may help slow climate change. Yet, incredibly,

Schwartz and Randall fail to mention the option at the heart of the problem—reducing the use of fossil fuels!

In 2004 the Hollywood disaster movie *The Day after Tomorrow* also imagined the consequences of a possible shutdown of the Gulf Stream. For dramatic effect the time-lines for the collapse are greatly compressed, and the changes are far grander than even those imagined in the Pentagon report. Scientists meanwhile have been working at understanding the consequences of a Gulf Stream collapse for biodiversity as a whole, and they *are* catastrophic. Biological productivity in the north Atlantic will fall by 50 per cent, and oceanic productivity worldwide will plummet by over 20 per cent.[228]

So what are the chances of the Gulf Stream shutting down this century? Under what conditions might it occur; and what would be the warning signs?

Although sailors have known of the Gulf Stream since the time of Columbus, the first map of it was not produced until Benjamin Franklin printed one in 1770. Today we know that it is the fastest ocean current in the world, and that it is complex, spreading out into a series of gyres and sub-currents as its waters move northward. The volume of water in its flow is simply stupendous. You will recall that ocean currents are measured in Sverdrups, and one Sverdrup equals a flow of 1 million cubic metres of water per second per square kilometre. Off Cape Hatteras, where the Gulf Stream angles off the coast towards deeper water, its flow reaches 87 Sverdrups, while at its peak, around 65 degrees of longitude West, the Gulf Stream flows at a rate of 150 Sverdrups. Overall, its flow rate is around 100 Sverdrups, which is 100 times as great as that of the Amazon.[137]

In its northern section the Gulf Stream is far warmer than the waters that surround it. Between the Faeroes and Great Britain, for example, the Gulf Stream is a balmy 8°C, yet the surrounding waters are at zero. The source of the Gulf Stream's heat is the tropical sunlight falling on the mid-Atlantic, and the current is a highly efficient way to transport it, for as Alfred Russel Wallace noted in 1903: 'As air is 770 times as light as water, it follows that the heat from one cubic foot of water will warm more than 3000 cubic feet of air.'[5] In the north Atlantic, where the Gulf Stream releases its heat, it warms Europe's climate as much as if the continent's sunlight were increased by a third.

And as the waters of the Gulf Stream yield their heat they sink, forming a great mid-oceanic waterfall. This waterfall is the powerhouse, as well as the Achilles heel, of the ocean currents of the entire planet, for history shows us that it has been interrupted time and again.

As the Earth's climate shifted from full ice-house mode 20,000 years ago to the mild climate of today, the Gulf Stream was repeatedly destabilised—most spectacularly between 12,700 and 11,700 years ago, when winter temperatures in the Netherlands plunged below −20°C and summer temperatures averaged just 13°C to 14°C. Between 8200 and 7800 years ago there was another collapse, while between 4200 and 3900 years ago it may have slowed again. On the two earliest occasions the disruption was caused by vast influxes of fresh water into the north Atlantic: the first by the bursting of an ice-dammed lake (of which the Great Lakes are a remnant) and redirection of melt-water from the Mississippi to the St Lawrence River; and then with the implosion of the remains of North America's Laurentide Ice Sheet and the draining of Lake Agassiz into Hudson Bay.[85] Fresh water disrupts the Gulf Stream because it dilutes its saltiness, preventing it from sinking and thus disrupting the circulation of the oceans worldwide.

The likelihood of the Gulf Stream slowing down again depends on whether a sufficient flow of fresh water can still be generated. Flows of a Sverdrup may have an effect, but several Sverdrups or more of freshwater flow is needed to seriously disrupt it. The frozen north contains enough ice to realise that liquid potential, and to this we must add the increasing rainfall already manifesting itself across the region.

From 1970, a steady freshening of the surface waters of the northeast Atlantic has been recorded: the salinity graph describes a graceful, downward arc that speaks powerfully of the emerging trend. Three decades ago the current's salinity median was 34,960 parts per million, but by 2000 it had fallen close to 34,900. In Denmark Strait the decline has been greater; from 34,920 parts per million to 33,870, though here the graph reveals a series of bumps and dips testifying to the influence of local freshwater flows. The average salinity of sea water is around 33,000 parts per million, so even such small changes are cause for concern, for it is the differential in salt content—at present just 1900 parts per million—that keeps the Gulf Stream moving.

Evidence of more widespread changes in the Atlantic was reported in 2003 by Ruth Curry of Woods Hole and her colleagues.[149] They carried out an exhaustive study that examined the salinity of the Atlantic Ocean from Pole to Pole over two fourteen-year periods, 1955–69 and 1985–99. They discovered changes 'of a remarkable amplitude', indicating that 'fresh water has been lost from low latitudes and added at high latitudes, at a pace exceeding the ocean circulation's ability to compensate'.[149] In other words, at all depths the tropical Atlantic is becoming saltier, while the north and south polar Atlantic are becoming fresher. The change, the researchers reasoned, was due to increased evaporation near the equator and enhanced rainfall near the Poles. When they found similar changes in other oceans, they realised that something—most probably climate change—had accelerated the world's evaporation and precipitation rates by 5 to 10 per cent.

This remarkable discovery has even greater potential impact on the Gulf Stream. The increasing tropical saltiness, the researchers suggest, will lead to a temporary quickening of the Gulf Stream that will paradoxically herald its abrupt shutdown. This will occur because of the extra heat transferred to the Poles, which will melt more ice and thus freshen the north Atlantic until the required Sverdrups flow into it, collapsing the system altogether.[150]

The Gulf Stream is just one part of a system of globe-circulating ocean currents, and researchers have also been observing changes elsewhere.

In early 2004 researchers from the CSIRO in Australia announced that they had detected a decrease in oxygen levels of around 3 per cent in deep subantarctic waters. Our knowledge of the variability in oxygen levels in the ocean depths over time is still slender, and several factors may explain the decrease (a phytoplankton bloom that sank to the bottom and rotted is one explanation); yet the figures worry some climate researchers because such a drop in oxygen is just what one might expect if the ocean's thermohaline circulation was slowing down, preventing oxygen from the upper layers of the ocean from mixing with the deep waters.

If the Gulf Stream were to fail, how fast might it happen? Ice-cores from Greenland indicate that, as the Gulf Stream slowed in the past, the island experienced a massive 10°C drop in temperature in as little as a decade.[124] Presumably, similarly rapid changes were also felt over Europe, although no fine-grained record of climate has survived to tell of it. So it is conceivable that extreme changes could be felt over Europe and North America within a couple of winters were the Gulf Stream to slow. It is even possible that the climatic see-sawing that happened at the end of the ice age may recur.

When is such an event likely to happen? Given uncertainty about the rate of melting of ice caps, and the complexity of other factors, it is difficult to be precise. Some eminent climatologists think that they are already seeing signs of a prelude to a shutdown. Were I forced to guess, I'd say that by 2080 Greenland may be 4°C warmer than today, which would melt enough ice to raise sea level by five centimetres; which may provide sufficient Sverdrups to shut the current down for some centuries. But with the melting of Greenland's ice halted by the cold conditions, the current would eventually re-start, and with that the melting of the ice would also recommence, initiating a see-saw pattern that could continue until the ice reserve reached a threshold, upon which it would have insufficient flow to disrupt the Gulf Stream. Not all agree, however, that a collapse or even a slowing of the Gulf Stream is imminent. Scientists at the Hadley Centre in England rate the chance of major disruption to the

Gulf Stream this century at 5 per cent or less. Their main concern, as far as abrupt shifts go, is an event that, while less widely known, could be even more catastrophic than Gulf Stream disruption—the collapse of the Amazon's rainforests.

SCENARIO 2: COLLAPSE OF THE AMAZON RAINFORESTS

In the 1990s the Hadley scientists used a global circulation model called HadCM3LC, which was the first to incorporate both the carbon cycle and a model of the Earth's major vegetation communities. In using this powerful new tool, researchers came up with astonishing results that underlined the importance of positive feedback loops.[136]

The most significant aspect of the carbon cycle, as manifested in the model, is the reserve of carbon in the soil, for that is a potential source of carbon dioxide so enormous that it dwarfs the amount stored in living vegetation. And the carbon in this store is finely balanced, with only a small change in temperature capable of turning soils from an absorber of CO_2 to a large-scale emitter. This change is brought about by bacterial decomposition: at lower temperatures it is slow, allowing carbon to accumulate, but as the soil warms decomposition accelerates and CO_2 is released in prodigious volume.[190] This is a textbook example of a positive feedback, where increasing temperature leads directly to a vast increase in CO_2 in the atmosphere.

The Hadley Centre's vegetation model, known somewhat whimsically as TRIFFID (Top-down Representation of Interactive Foliage and Flora Including Dynamics), is as yet a crude approximation of reality, for it admits only five plant categories: broadleaf trees, needle-leaf trees, the two principal kinds of grasses (C3 and C4), and shrubs. Nonetheless these categories comprise Earth's broad functional vegetation types. As the concentration of atmospheric CO_2 is increased in their virtual world, the plants—particularly in the Amazon—start behaving in unusual ways.

The problem begins with the nature of rainfall over the Amazon. Recall the isotopes of oxygen: ^{16}O and ^{18}O. Of the two, ^{16}O is lighter, and because of that is preferentially taken up by evaporation. When scientists examined the water falling in the western Amazon Basin, they discovered that it was very low in ^{18}O. That is because it had been recycled into the atmosphere so many times that most of the ^{18}O had been left behind far to the east. This tells us that the plants of the Amazon effectively create their own rainfall, for so vast is the volume of water transpired by them that it forms clouds that are blown ever westwards where the moisture falls as rain, only to be transpired again and again.

Transpiration is vital to rainfall in the Amazonian rainforest, and it turns out that CO_2 does odd things to plant transpiration. Plants, of course, generally don't wish to lose their water vapour, as they have gone to some trouble to convey it from their roots to their leaves. But inevitably they do lose some whenever they open the breathing holes in their leaves (stomata). Their main purpose in doing this is to gain CO_2 from the atmosphere, and they will keep their stomata open only as long as required. Thus, as CO_2 levels increase, the plants of the Amazonian rainforest will keep their stomata closed for longer, and transpiration will be reduced. And with less transpiration there will be less rain.

TRIFFID indicates that by around 2100, levels of CO_2 will have increased to the point that Amazonian rainfall will reduce dramatically, with 20 per cent of that decline attributable to closed stomata. The rest of the decline, the model predicts, will be due to a persistent El Niño-like climate that will develop as our globe warms.

Incidentally, another positive feedback loop will kick in at this stage, for research into the impact of El Niño on carbon sequestration has revealed that it transforms the world's landmasses from carbon sinks into carbon sources which have, on average, increased the accumulation of CO_2 in the atmosphere by 0.6 parts per million.[189]

The cumulative impact of all of these changes is to reduce rainfall from the current basin-wide average of 5 millimetres per day to 2 millimetres per day by 2100, while in northeastern Amazonia it will fall to almost zero.[135] These conditions, combined with a basin-wide rise in temperature of 5.5°C will, the model indicates, stress plants to the point that collapse of the Amazonian rainforest will become inevitable. With the loss of the rainforest canopy, soils will heat and soil decomposition will proceed at an even more rapid rate, which will lead to the release of yet more CO_2. This constitutes a massive disruption of the carbon cycle, reducing the storage of carbon in living vegetation by 35 gigatonnes, and soil carbon storage by 150 gigatonnes.[135] These are huge figures—totalling around 8 per cent of all carbon stored in the world's vegetation and soils!

The ultimate outcome of this series of positive feedback loops is that by 2100 the Earth's atmosphere will have close to 1000 parts per million of CO_2 rather than the 710 predicted in earlier models.[190] Surface temperatures in the Amazon will rise by 10°C rather than the 5.5°C predicted, rainfall in the basin will drop by 64 per cent, there will be a 78 per cent loss of carbon stored in vegetation, and a 72 per cent loss of soil carbon.[134]

One of the most frightening aspects of this modelling experiment is what remains in the Amazon Basin after the change. Most of the tree-cover is replaced by grasses, shrubs, or at best a savannah studded with the odd tree. Large areas, however, become so hot and blighted that they cannot support even this reduced vegetation, and so turn into a barren desert.[135] Yet the team at Hadley remain somewhat optimistic about the fate of these regions, for while TRIFFID cannot find plants that will grow there, the scientists believe that 'even at mean annual temperatures of approaching 40°C a sparse cover of semi-desert plants might be possible'.[134]

When might all of this happen? If the model is correct we should start to see signs of rainforest collapse around 2040, and the process should be complete this century, by which time rainforest cover will have been reduced from its present 80 per cent to less than 10 per cent. Half of the deforested region will turn to grass, the other half to desert.[135] Other rainforests round the world may by then be showing signs of similar stress, for most depend to some extent on transpired water for rain.

What is so terrifying about this scenario is that it will greatly hasten climate change, making many of its most pernicious consequences inevitable.

The Arctic Ocean is where manifestations of the third of the possible great shifts may first appear. In this particular scenario, the cause is something that has not yet figured large in the workings of the climate modellers, but which prehistory teaches us should be given serious attention: a sudden unleashing of the clathrates.

SCENARIO 3: METHANE RELEASE FROM THE SEA FLOOR

Clathrates is Latin for 'caged' and the name refers to the structure of this ice-methane combination in which ice crystals trap molecules of methane in tiny 'cages'. Clathrates are also known as the 'ice that burns'. They contain lots of gas under high pressure, which is why pieces of the icy substance hiss, pop and, if lighted, burn when brought to the surface.[220]

Massive volumes of clathrates lie buried in the seabed right round the world—perhaps twice as much in energy terms as all other fossil fuels combined. Optimum conditions for clathrate formation exist where ocean water is more than 400 metres deep, and bottom temperatures are below 1–2°C. The material is kept solid only by the pressure of the overlying water and the cold. While most clathrates lie kilometres below the sea's surface, very large volumes can be found in the Arctic Ocean, for there temperatures are sufficiently low, even near the surface, to keep them stable.

It's illustrative of the endless ingenuity of life that some marine worms survive by feeding on the methane in clathrates. They live in burrows within the icy matrix, which they 'mine' for their energy requirements, and as there are between 10,000 and 42,000 trillion cubic metres of the stuff scattered around the ocean floor (which compares favourably with the 368 trillion cubic metres of recoverable natural gas in the world) it's not surprising that both worms and the fossil fuel industry can see a future in this paradoxical material.[179]

If pressure on the clathrates was ever relieved, or the temperature of the deep oceans were to increase, colossal amounts of methane could be released. We have seen the consequences of one such release in the North Sea 55 million years ago, but palaeontologists are now beginning to suspect that the unleashing of the clathrates may have been responsible for a far more profound change—the biggest extinction event of all time.

Two hundred and forty-five million years ago around nine out of ten species living on Earth became extinct. Known as the Permo-Triassic extinction event, it carried off an early radiation of mammal-like creatures, thus opening the way for the dominance of the dinosaurs—which seems ironic given that another similar event could destroy the civilisation of the most successful mammalian species yet.

The cause of the extinction is hotly debated, though there are two front runners: the collision of an asteroid with Earth, and a massive outpouring from the Siberian Trap volcanoes which released up to 2 million cubic kilometres of lava and billions of tonnes of CO_2 and sulphur dioxide. It's this second hypothesis that is growing in strength, and it's the way that these volcanic gases are thought to have interacted with clathrates that merits our attention here.

So vast was this input of greenhouse gas to the atmosphere that it was thought to have led to an initial rise in global average temperature of about 6°C. This co-occurred with widespread acid rain caused by the sulphur dioxide, which released yet more carbon. Such was the total impact of the increasing temperature thereby generated that it triggered the release of huge volumes of methane from the tundra and clathrates on the sea floor.[204]

Incidentally, while we're looking at events under the ocean, the methane and CO_2 stored in the permafrost must not be forgotten. Massive amounts of these gases are locked away in the perennially frozen soil, and are even more likely to be released by climate change than the clathrates. One intriguing thing about the atmosphere around the time of the extinction was its low oxygen content. Two hundred and eighty million years ago it stood at 21 per cent (the same as today), yet by 260 million years ago it had fallen to 15 per cent, and then to only 10 per cent at the time of the Permo-Triassic extinction. This, at least one researcher thinks, may have been caused by the sudden release of methane, for the gas would have quickly oxidised to CO_2 and water in the atmosphere, and in so doing soaked up vast quantities of the free atmospheric oxygen.[205]

Clathrates are important structural elements in sea floor stability, and their sudden sublimation could lead to 'slumping' and the generation of tsunamis of unprecedented power. Indeed, a slump off the Carolina coast 15,000 years ago is thought to have released enough methane to increase atmospheric concentrations by 4 per cent.[220] It's sobering to ponder that there could be an unstable clathrate time bomb off a beach near you!

Of the three scenarios presented, the release of the clathrates is the least likely to occur this century. Only a massive warming event, one suspects, could trigger it.

The shutting down of the Gulf Stream is unique among the possibilities canvassed here in that it is a powerful negative feedback loop which for the north Atlantic rim countries at least—and perhaps for the planet as a whole—will temporarily and dramatically reverse the warming trend. Thus, from a Gaian perspective, disrupting the Gulf Stream is akin to cutting off a gangrenous limb before it corrupts the entire body. The other two scenarios, in contrast, are positive feedback loops, one of which is the most powerful in Earth's history. When thinking about these potential catastrophes, it's important to realise that, as when firing a gun, the possibility of human control is only there at the very beginning of the process—before we trip the trigger.

There's one other feedback loop that I'd like to mention here, not because it's on a scale commensurate with the three great phase shifts discussed above, but because it concerns ourselves, is already occurring, and may be the trigger for further change.

Throughout our history we have engaged in a constant battle to maintain thermal comfort, which has been very costly in terms of time and energy. Just think of the hundreds of slight shifts in body position and posture we undertake every day and night—the taking off and putting on of overcoats, hats, etc.—that are simple manifestations of that battle. Indeed, purchasing a house—our greatest personal expense—is primarily about regulating our local climate. Today we can use fossil fuels to heat and cool these environments, an enterprise costly both in terms of energy use and the environment. In the US 55 per cent of the total domestic energy budget is devoted to home heating and air conditioning—while home heating alone costs Americans $44 billion per year.[199]

As our world becomes more uncomfortable courtesy of climate change, it is inconceivable that the demand for air conditioning will lessen. In fact, during heatwaves it could mean the difference between life and death. But, unless we change our ways, that demand will be met by burning fossil fuels, which represents a very powerful positive feedback loop. An insatiable demand for air conditioners is already evident in countries such as the US and Australia where, until recently, construction codes for houses have been appallingly lax in regard to energy use. A situation may even develop whereby, in order to cool our homes, we end up cooking our planet.

TWENTY-TWO

CIVILISATION: OUT WITH A WHIMPER?

> Unless we stop now, we will really doom the lives of our descendants. If we just go on for another forty or fifty years faffing around, they'll have no chance at all, it'll be back to the stone age. There'll be people around still. But civilisation will go.
>
> James Lovelock, *Independent*, 24 May 2004.

Our civilisation is built on two foundations: our ability to grow enough food to support a large number of people who are engaged in tasks other than growing food; and our ability to live in groups large enough to support great institutions. These clusters are known as cities, and it is from their inhabitants, the citizens, that the word civilisation is derived.

Today, very large cities lie at the heart of our global society, and our most valued institutions are found in them. Unless they are subsidised from outside, population centres with fewer than 10,000 inhabitants are unlikely to possess the full spectrum of general health care; while those with only 100,000 generally lack a quality centre of higher learning and an orchestra. Even small cities—those with around a million people—may lack an opera house, a world-class museum and some specialist medical care. And there is a dramatic difference in the job opportunities—particularly in specialist fields—available in a city of 5 million as opposed to a city of 1 million.

Cities are central to civilisation, and yet they are fragile entities vulnerable to the stresses brought about by climate change. It is therefore important to consider cities in relation to the provision of their basic needs—food, water and power.

The only other creatures to have achieved anything like a city are the social insects, and so small are their bodies and their demands for resources that a few hectares of habitat is all that is required to satisfy their needs. Our cultures, in contrast, straddle continents, and our cities have become rainforest-like in their complexity. In cities almost every job is specialised: being a mere 'secretary' will no longer do—one must instead be a legal or medical secretary or some such. And a doctor is best served not by becoming simply a general practitioner but a sports specialist, a proctologist, or an expert on aged care. This is the equivalent, in human terms, of being a matanim cuscus or golden toad—and in the natural world such species are seen only in rainforests, because only there is the supply of energy and moisture sufficiently large and regular to feed such complex and large aggregations of life.

As we have seen, if we cut off the water or sunlight to a rainforest for even a brief period it is liable to collapse and its specialist species become extinct. Climate change has done just that in areas of Costa Rica and Papua New Guinea, and it is predicted to do so in regions such as the Amazon. Now, let's perform a mental experiment. Think of a city you are familiar with and imagine what it would be like if its citizens woke one morning to discover that no fresh water issued forth from their taps.

No clothes could be washed, no toilets would flush, filth would accumulate and people would become thirsty very quickly. And imagine the result if petrol supplies came to a halt. Food could not be delivered, garbage wouldn't be removed, and people couldn't get to work.

Could climate change threaten the resources required by cities to survive? Physicist Stephen Hawking has said that within a thousand years increasing CO_2 will boil the surface of our planet and humans will need to seek refuge elsewhere. This is an extreme opinion. More mainstream are the views of Jared Diamond, who has examined civilisations that have collapsed in the past.[223] He has found that exhaustion of the resource base is a key reason why even large, complex, literate societies such as the Maya failed. In its effect, a rapid shift to another kind of climate could place a similar stress on our global society, for it would alter the location of sources of water and food, as well as their volume.

Humans seem to be eternally optimistic about their capacity to adapt, and in the face of such a possibility those I have spoken to have suggested deriving water from hydrogen power plants, towing icebergs, or growing crops hydroponically as solutions. All of these responses may assist a privileged few, but so enormous is the problem and so long would it take to ramp up any such solutions to a global scale, that in the face of swift climate change they offer no hope for the great majority of us.

The threat posed by increased climate variability is a very real one. A good example of the relationship between climate variability and human population size is provided by Australia. It is unique among the larger nations in consisting of either very small settlements or large cities, for the middle-sized towns that predominate elsewhere in the world are almost entirely absent. This is a consequence of the cycle of drought and flood that has characterised the land from first settlement.

Small regional population centres have survived because they can batten down the hatches and endure drought, and large cities have also survived because they are integrated into the global economy. The resource networks of towns, however, are smaller than the region affected by climate variability, making them vulnerable to swings in income. Typically what happens is that, as a drought progresses, the farm machinery dealership and automotive dealership closes down. Then, with everyone feeling the pinch the pharmacist, the bookseller and the banks leave. When the drought finally breaks and people have cash again, these businesses do not return, for instead people travel to larger centres to buy what they need, and in time end up moving there.

The Australian example shows that climate variability has in fact encouraged the formation of cities: today it is the most urbanised nation on Earth. But the only reason that Australia's cities are refuges from climate variability is that they draw their resources from a region broader than that affected by the continent's droughts and floods. But with climate change we are talking of a global phenomenon: the entire Earth will be affected by climate shifts and extreme weather events of ever greater amplitude.

Water will be the first of the critical resources to be affected, for it is heavy, commands a low price and thus is unprofitable to transport long distances. This means that most cities source their supply locally, in areas small enough for mild climate change to have an impact. Already we have seen how Perth and Sydney sit on the knife-edge in terms of their water supply, and doubtless more cities will join the list as water shortages increase around the world. Food such as grains, in contrast, is easily transported and is often sourced from afar, which means that only truly global disruptions would cause shortages in the world's cities.

To date, impacts from climate change have been relatively small. For the past eight years, droughts and unusually hot summers have caused world grain yields to fall or stagnate, during which time the number of extra mouths humanity must feed has grown by 600 million. The peak in cereal reserves, of around 100 days, was reached in 1986, and fell to a low of fifty-five days in 1995. Although substantial wheat surpluses were recorded in 1999 and 2004, overall the trend in world food security has been a downward one.

When it comes to climate change, cities are more like plants than animals, for they are immobile, and depend on intricate networks to supply the water, food and energy that they require. We should worry that whole forests of trees are already dying as a result of climate change, for cities will likewise begin to die when this phenomenon overwhelms the capacity of their networks to supply the essentials. This may occur through the repeated battering of extreme weather events, rising seas and storm surges, extreme cold or heat, water deprivation or flood, or even disease.

It's worth diverting here from our broader discussion of cities, to review the idea, raised by the American coal industry, that increasing levels of CO_2 will 'fertilise' the world's crop plants, providing a solution to world hunger. Many experiments in which plants have been fed artificially high levels of CO_2 have now been completed, and botanists Elizabeth Tansley and Stephen Long have analysed the results.[215]

It turns out that trees benefit far more than shrubs or grasses from increases in CO_2, and that the species that benefit least are grasses belonging to a group that includes our most important crops. Rice, for example, showed an increase in yield of only 6 per cent in response to a doubling of CO_2, while wheat yields rose by only 8 per cent. In future, crops will be stressed by higher temperature, more ozone at ground level and changes in soil moisture, all of which will decrease yields. Thus, rather than an agricultural paradise, a CO_2-rich world promises to be one in which crop production is lower than today.

On pondering how few crop species support us, the philosopher Ronald Wright remarked 'we have grown as specialised, and therefore as vulnerable, as a sabre-toothed cat'.[214] It is often said that farmers will adapt by planting new crops suited to the new climate, if any can be found. But one of the worrying things about climate change is that the overall biological productivity of our planet is decreasing: in other words there's less cake to go around.

Because of the differing capacities of rich and poor, and of human versus natural systems to adapt to climate change, some in the environment movement are characterising adaptation as having acquired 'a genocidal meaning'.[197] By this they mean that a cosseted, wealthy few may survive climate change by retreating to some refuge, but the vast majority will inevitably perish, as will the bulk of Earth's species and ecosystems.

English environmental politician Aubrey Meyer pointed out how this matter is being discussed at the highest levels. Economists who participated in the IPCC discussions stated that doing anything serious about climate change was too expensive to be worthwhile, leading in Meyer's view to 'the effective murder of members of the world's poorer populations', and whose lives by the economists' estimates were worth only a fifteenth that of a rich person.[197] I agree with Meyer that adaptation of this sort is genocide, and attempted Gaia-cide as well. For this reason I believe that our efforts must be put into avoiding the change in the first place.

So, could the day ever come when the taps run dry, or power, food or fuel is no longer available in many of the world's cities? That depends on the amplitude of climatic change that increased greenhouse gases bring: if it exceeds the extent of a city's resource procurement network, then collapse is inevitable. We have no figures as to what degree of warming could trigger such a collapse, but 0.63°C of warming has proved sufficient to inflict acute distress on large regions such as the Sahel, the Arctic and subantarctic waters. Three degrees of warming—five times as much as has been experienced thus far—will have more wide-reaching impacts: enough perhaps to destabilise continent-sized regions. At the highest end of the scale—11°C of warming—the impacts are unimaginable, and would threaten our species as a whole.

Threats to our civilisation from declining rainfall and food shortages are only those that may result from a continuation of current trends. If we were to experience an abrupt climate shift, it is possible that a near-eternal, dreary winter would descend on the cities of Europe and eastern North America, killing crops, and freezing ports, roads and human bodies as well. Or perhaps extreme heat, brought on by a vast exhalation of CO_2 or methane, will destroy the productivity of oceans and land alike. Given the scale of the change confronting us, I think that there is abundant evidence to support Lovelock's idea that climate change may well, by destroying our cities, bring about the end of our civilisation.

Humanity, of course, would survive such a collapse, for people will persist in smaller, more robust communities such as villages and farms—situations that resemble temperate deciduous forests rather than rainforests. Towns have relatively few people just as temperate forests have relatively few species, and the inhabitants of both are hardy and multi-skilled. Just think of the maple with its skeletal winter form and verdant summer manifestation, or the country house with its own water tank and vegetable garden. These characteristics mean that both the maple and the rural family can withstand periods of shortage that would destroy a city or a rainforest.

For a small town a drought may be a worry, but because whatever rain falls on a non-porous roof is captured in a water tank, it will benefit from even the briefest shower. Dams, in contrast, require substantial rain to create flows, because lots of water soaks into the soil. Likewise a late fuel delivery or failure of power supply is annoying to those in small communities, but the impact on them is nothing compared with the dilemma faced by the inhabitants of a city tower block. In the long term, however, even medium-sized towns lack the know-how to keep their complex infrastructure—such as medical services and machinery—working. They are ultimately just as dependent on our civilisation as are the city dwellers, meaning that a climate-change-driven dark ages will affect them as well.

We have seen that human health, water and food security are now under threat from the modest amount of climate change that has already occurred. If humans pursue a business-as-usual course for the first half of this century, I believe the collapse of civilisation due to climate change becomes inevitable.

We have known for some decades that the climate change we are creating for the twenty-first century was of a similar magnitude to that seen at the end of the last ice age, but that it was occurring thirty times faster. We have known that the Gulf Stream shut down on at least three occasions at the end of the last ice age, that sea levels rose by at least 100 metres, that the Earth's biosphere was profoundly reorganised, and we have known that agriculture was impossible before the Long Summer of 10,000 years ago. And so there has been little reason for our blindness, except perhaps for an unwillingness to look such horror in the face and say, 'You are my creation.'

4

PEOPLE IN GREENHOUSES

TWENTY-THREE

A CLOSE-RUN THING

> If it had turned out that chlorine behaved
> chemically like bromine, the ozone hole would by the 1970s
> have been a global, year-round phenomenon, not just an
> event of the Antarctic spring. More by luck than by wisdom,
> this catastrophic situation did not develop.
> Paul Crutzen, *Nature*, 2002.

Throughout 2004 humanity seemed to be in a hopeless muddle in developing a response to the climate change crisis. The fate of the Kyoto Protocol hung in the balance as Russia decided whether to ratify it or not (it did), and the US and Australia hardened their opposition to the agreement. It was a depressing time to be writing a book such as this. Then I discovered that around twenty years ago the world had performed a full dress rehearsal for Kyoto, stumbling blocks and all. It was called the

Montreal Protocol, and it sought to limit the emission of chlorofluoro-carbons (CFCs), which destroy ozone.

Before we contemplate the global response to climate change, it's worth having a look at CFCs and the international agreement that dealt with the threat they represented to life on Earth.

A particular form of oxygen, ozone was discovered in the laboratory in the 1830s, and by 1850 it was detected occurring naturally in the atmosphere. Throughout the nineteenth century measurements at ground level were being undertaken around Europe, and it is interesting to note that the 1873 levels recorded in Paris were roughly half that of today. This is symptomatic of a global increase in ozone at ground level, where it is a serious toxic pollutant.

By the 1920s Oxford University's Gordon Dobson and his collaborator F. A. Lindeman (later Lord Cherwell) realised that ozone played an important role in the stratosphere, and to this day the amount of ozone in the atmosphere is measured in 'Dobson units'. In 1948 an International Ozone Commission was set up to study the gas. Thus far, ozone research had been driven by pure scientific curiosity, for no one had any idea that it might affect the future of humanity. Then in 1957—the so-called International Geophysical Year, when governments around the world spent a billion dollars to better understand Earth processes—a sustained effort to measure ozone began.

The first signs of a problem arose in the 1970s, when readings of ozone concentration in the stratosphere above the Antarctic began to look decidedly odd. The instruments were reporting ozone loss at a phenomenal rate: in 1955 the air over the Antarctic held 320 Dobson units. In 1975 there were 280 Dobson units, and by 1995 only 90. Given the relative stability of ozone levels being reported elsewhere, the readings seemed so bizarre that for a vital decade they were put down to some sort of instrumental error. Yet as early as 1974 three scientists—Paul Crutzen, F. Sherwood Rowland and Mario Molina—were arguing that the depletion was real, and that the cause was manmade chemicals. In 1995 the three were awarded the Nobel prize in chemistry for this pioneering work.

When first reported in the press the 'hole in the ozone layer' was at times portrayed with humour—as if scientists had become Chicken Little warning that the sky was falling. Even Sherwood Rowland reacted with an air of unreality to his grim research findings. He later recalled, 'I came home one night and told my wife, "The work is going very well but it looks like the end of the world."'

A 'hole' in the ozone layer is defined as an area of atmosphere with less than 220 Dobson units of ozone. By 2000 the hole had become a chasm spanning 28 million square kilometres, and around it had spread a halo of thinned ozone covering most of the globe below 40°S. By the 1990s a second hole had appeared, this time over the Arctic.[62] Even over the tropics, ozone concentration was reduced by around 7 per cent.

So what exactly is ozone, and why is it important? The oxygen that keeps your body alive consists of two atoms of oxygen joined together, but high in the stratosphere, ten to fifty kilometres above our heads, ultraviolet radiation occasionally forces an extra oxygen atom to join the duo. The result is three-atom molecules of a sky-blue gas known as ozone. Ozone is unstable, for it is constantly losing its additional atom, but new trios are forever being created by sunlight, so a constant amount persists—about ten parts per million (one of every 100,000 molecules) in an undamaged stratosphere. Ozone is six times as abundant in the stratosphere as it is at sea level, yet if all the planet's stratospheric ozone were brought down to sea level, it would form a layer just three millimetres thick.

If the great aerial ocean is Earth's blood supply, then ozone is its sunscreen. Two-atom oxygen is able to block ultraviolet (UV) radiation that comes in wavelengths shorter than 0.28 microns, but ozone can block UV wavelengths between 0.28 and 0.32 microns. It shields us from around 95 per cent of the UV radiation (that is, radiation at wavelengths shorter than 0.4 microns) reaching Earth. Without ozone's very high sun-protection factor, ultraviolet radiation would kill you fast, by tearing apart your DNA and breaking other chemical bonds within your cells.

The destruction of the ozone layer began long before anyone was aware of it. Fluorocarbons (CFCs and HFCs) had been invented by industrial chemists in 1928, and were found to be very useful for refrigeration, making styrofoam, as propellants in spraycans, and in air conditioning units. Their remarkable chemical stability (they do not react with other substances) made people confident that there would be few environmental side effects, so they were embraced by industry.

By 1975 spraycans alone were flinging 500,000 tonnes of the stuff into the atmosphere, and by 1985 global use of the main types of CFCs stood at 1,800,000 tonnes. It was their stability, however, that was a key factor in the damage they caused, for they lasted a very long time in the atmosphere.

CFCs evaporate easily, and once released into the great aerial ocean it takes about five years for air currents to waft them into the stratosphere, where UV radiation slowly breaks them down, causing the release of their chlorine atom. It is the chlorine in CFCs that is so destructive to ozone—just a single atom can destroy 100,000 ozone molecules—and its destructive capacities are maximised at temperatures below −43°C. This is why the ozone hole first emerged over the South Pole, where the stratosphere is a frigid −62°C. At −42°C the stratosphere over the North Pole is balmy by comparison, and it took longer for the chlorine there to deplete ozone to the point that a 'hole' formed.

It was James Lovelock—originator of the Gaia hypothesis, but at this time a self-employed scientist working independently of institutions—who invented the machine used to detect CFCs in the atmosphere. Because he could not get funding for the project, he made it out of spare parts in his garage, then took his contraption on an Antarctic cruise. Despite extensive measurements, Lovelock found such minuscule amounts of the chemicals in the atmosphere that at first he thought his work to be useless. It was only in 1973, as a result of a chance meeting with a Dr Machta, during a coffee break at a conference, that the true significance of his findings was revealed.

Dr Machta was a chemist who worked for DuPont, the company that made most CFCs, and a quick calculation revealed that, tiny though the total concentration Lovelock detected was, it amounted to almost all of the CFCs ever made. The stuff simply did not go away, and that was enough to have Dr Machta discussing Lovelock's findings with other chemists, including Dr Mario Molina, who discovered the link between CFCs and ozone.[230]

Molina discovered that CFCs had elevated chlorine levels in the stratosphere to five times their historic background level. This was bad enough, but it is a matter of dumb luck that our world did not enter a far more severe environmental crisis—perhaps one leading to the collapse of societies—some thirty years ago. This could have occurred if industrial chemists had used bromine instead of chlorine.

Bromine and chlorine can be used interchangeably for many purposes, and the fact that chlorine is used more often is largely the result of economics, for bromine is somewhat more expensive (and more reactive) than chlorine, a situation made worse by the fact that you get somewhat less fluorocarbon per gram of bromine-based product.[185]

Although bromine lasts just one year in the stratosphere as compared with chlorine's five, bromine is forty-five times more effective in destroying ozone than chlorine, and so swiftly would it have torn asunder those precious 10 parts per million of ozone that Earth's sunscreen would have been destroyed even before Sherwood Rowland had made his Nobel-winning discovery. Just how close the world came to such a fate can be seen in the uses to which industrial chemists had already put bromine.

In the 1980s (pause a bit) bromotrifluoromethane and bromochlorodifluoromethane—their trade names Halon-1301 and Halon-1211 respectively—came into widespread use in fire-suppression systems, particularly in art galleries and museums where using water might cause damage. Because these chemicals are ten times as potent in destroying ozone as CFCs, they were banned under the Montreal Protocol, but bromine is still released into the atmosphere as a result of human activities, mostly from the use of agricultural pesticides.

So what might have happened if BFCs, rather than CFCs, had found favour with industry? A hint of the damage that may have occurred can be garnered from the job now being done by CFCs. As a result of the hole they punched in the ozone layer, people living south of 40° are experiencing a spectacular rise in the incidence of skin cancer. At 53°S, Punta Arenas in Chile is the southernmost town on Earth. Since 1994 skin cancer rates there have soared by 66 per cent. Even at lower latitudes—and closer to the great centres of human population—shifts in cancer rates are evident. In America, for example, the chance of getting melanoma was one in 250 just twenty-five years ago. Today it is one in 84. Ultraviolet radiation also causes eye damage, and its incidence is also rising. Researchers estimate that humans—and anything else with eyes—will experience a 0.5 per cent increase in cataracts for every 1 per cent decrease in ozone concentration. As 20 per cent of cataracts are due to UV damage, the rate of cataract-caused blindness looks set to rise fast, especially among those who lack the means to protect themselves. A third major impact on human health comes from the capacity of UV to damage the immune system. This will manifest itself as a general sickening in the stricken communities. Among vulnerable groups such as the Inuit, these impacts are already being felt.

It is not just human bodies that are affected by UV, for the impacts of its increase will be felt throughout the ecosystem. The microscopic single-celled plants that form the base of the ocean's food chain are severely affected by it, as are the larvae of many fish, from anchovy to mackerel. Indeed anything that spawns in the open is at risk, and a compelling new study shows how that risk is heightened (to 90 per cent mortality) if accompanied by rising sea temperatures and salinity.[225]

So vulnerable are many marine species that without stratospheric ozone they would go into a swift decline, precipitating a collapse of the oceans' ecosystems. We have already seen the exquisite vulnerability of amphibian larvae to increased UV. Their fate is merely an early symptom of what might have occurred on land, for all natural ecosystems are vulnerable. And nor would agriculture escape its effects. The yield of crops such as peas and beans, for example, decreases 1 per cent for every extra per cent of UV radiation received.

Had humans found bromine cheaper or more convenient to use than chlorine, it is quite likely that by the time Paul Crutzen and his colleagues made their discovery, we all would have been enduring unprecedented rates of cancer, blindness and a thousand other ailments, that our food supply would have collapsed, and that our civilisation itself was under intolerable stress. And we would have had no idea of the cause until it was too late to act.

For a decade after Crutzen and his team published their paper linking CFCs to ozone decline, the problem kept getting worse, yet scientists were unable to marshal proof positive that Crutzen's hunch was right. Such were the implications of ozone depletion, however, that colour images of the ozone hole shown on television screens around the world convinced thousands of people that action needed to be taken, even if only as a precaution. Politicians were bombarded with letters requesting the chemicals be banned. DuPont was the company responsible for most of their manufacture, and in retaliation they and other producers launched a massive public relations campaign aimed at discrediting the then-tenuous link between their products and the problem—and they had a point, for science was still unable to provide conclusive proof of CFCs' damaging impact.

Yet public feeling on this issue would not be appeased and, despite the howls of protest from industry about cost, representatives of twenty countries met in Vienna in 1985 and signed the Vienna Convention for the Protection of the Ozone Layer. Like today's Kyoto Protocol, the document was described as 'a toothless expression of hopes'.[70] By 1987, however, when scientific proof of the link between CFCs and ozone depletion was announced, it had given birth to the Montreal Protocol, in which the world's governments pledged to phase out the offending chemicals.

Today we know just how much was riding on the successful passage of the Montreal Protocol. Had it not been enacted, by 2050 the middle latitudes of the Northern Hemisphere (where most humans live) would have lost half of their UV protection, while equivalent latitudes in the Southern Hemisphere would have lost 70 per cent. As it was, by 2001, the Protocol had limited real damage to around a tenth of that.

Since its inception the Protocol has been tightened twice—in 1990 and 1992. And oddly enough, the reduction in CFCs was achieved at a net savings to the companies involved, and to the global economy. It may seem difficult to believe that government regulation could be good for the economy, but consider how Nortel, a US telecommunications company, benefited from the regulation. It had used the chemicals as cleaning agents and in the late 1980s it was forced to invest $1 million in new hardware; but once redesigned cleaning systems were in place and operating it saved $4 million in chemical waste disposal costs and CFC purchases.[209] Furthermore, the early adoption by the US of regulations to reduce the emission of CFCs gave American-based firms a head start on the rest of the world in innovating alternative chemicals.[236]

As with Kyoto, not all countries were initially bound by the Montreal Protocol. Indeed, China continues to produce CFCs, and may well go on polluting until 2010, when under the treaty it must cease. Despite such exceptions, the Montreal Protocol marks a signal moment in human societal development, for it represents the first ever victory by humanity over a global pollution problem. Today, there are stirrings of hope that we may have beaten that particular problem, for in 2004 the ozone hole over the Antarctic reduced by 20 per cent. Because the size of the hole waxes and wanes from year to year, we cannot be certain that this decrease signals the end of the problem. Nevertheless, scientists are optimistic that in fifty years' time the ozone layer will be returned to its former strength.

One could be forgiven for thinking that, with such a stunning all-round success to point at, the nations of Earth would have jumped at the chance to address climate change using a similar mechanism. And at first there was great enthusiasm for an international treaty to limit emissions of greenhouse gases. So what happened?

TWENTY-FOUR

THE ROAD TO KYOTO

> It is among those nations that claim to be
> the most civilised, those that profess to be guided by a knowl-
> edge of laws of nature, those that most glory in the advance
> of science, that we find the greatest apathy, the greatest
> recklessness, in continually rendering impure this all-impor-
> tant necessity of life…
>
> Alfred Russel Wallace, *Man's Place in the*
> *Universe*, 1903.

The Kyoto Protocol may be the most bitterly contested international treaty ever to be realised, which when one considers its modest goals appears strange indeed. Two big reasons for this are economics and politics.

In the developed world energy use is growing at the rate of 2 per cent per annum or less, and with such low rates of growth the only way for one sector (such as wind, gas or coal) to grow is to take from another sector's share. Kyoto will have a big influence on that outcome, and a furious struggle is ensuing between the potential winners and losers.

The treaty also marks a great divide, on one side of which stand those who are certain it is essential to Earth's survival, and on the other those who are fiercely opposed on economic and ideological grounds. Many in this group think Kyoto is economically flawed and politically unrealistic. Others believe that the entire climate change issue is hogwash.[125]

During its long gestation Kyoto was frequently wished or declared dead.[128] On 16 February 2005, however, ninety days after Russia ratified (bringing the number of ratifying nations to fifty-five, and the proportion of emissions from ratifying nations above 55 per cent) the Protocol came into force. The USA, Australia, Monaco and Liechtenstein remain outside it but, as with the creation of any large trading bloc, the pressure to join will now be incremental and unyielding. Kyoto is in its infant stages, but even now it's clear that it will influence all nations for decades to come.

The road to Kyoto began in 1985 with a scientific conference in Villach, Austria, which produced the first authoritative evaluation of the magnitude of climate change facing the world. It was followed in June 1988 by a meeting in Toronto attended by around 300 scientists and policy-makers from forty-eight countries. Although it possessed no special status, it quickly became known for its 'Call for Action' to reduce CO_2 emissions by 2005 to 20 per cent below those of 1988. No further global action was taken until 1992 when, at the Rio Earth Summit, 155 nations signed the UN Framework Convention on Climate Change, which designated 2000 as the year by which signatory nations would reduce their emissions to 1990 levels. This target was, in hindsight, wildly optimistic.

Following five years of protracted negotiations, on 11 December 1997 the signatories of the Framework Convention came to a new understanding as to how emissions would be reduced. Known as the Kyoto Protocol (because it was negotiated in the Japanese city of that name), it established two important things: the assignment of greenhouse emission targets to developed countries, and arrangements for emissions trading

224 PEOPLE IN GREENHOUSES

of the six most important greenhouse gases, a trade which is valued at US $10 billion.[133] With all countries signed up to the agreement, only ratification was required to bring it into effect.

Because CO_2 is by far the most significant of the greenhouse gases, you can think of Kyoto as setting national carbon budgets for the signatory nations, and establishing a new currency—a sort of 'carbon dollar', the trade of which allows industries to cost-effectively reduce emissions. It sounds like a sensible scheme, yet it would take until late 2004, six years after they agreed to it, for enough nations to ratify the treaty and propel it into life.

Perhaps the most damning criticism of Kyoto is that it is a toothless tiger. And that is indisputably true, for the momentum of climate change is now so great that Kyoto's target of reducing CO_2 emissions by 5.2 per cent is little more than irrelevant. Incidentally, those outside Kyoto are doing even worse: the US National Commission on Energy Policy suggests the adoption of a carbon trading scheme which would, in their words, 'go nowhere near those required for the United States under the Kyoto Protocol'.[226]

If we are to stabilise our climate, Kyoto's target needs to be strengthened twelve times over: cuts of 70 per cent by 2050 are required to keep atmospheric CO_2 at double the pre-industrial level. The Protocol's advocates, however, know how difficult it has been to get even a toothless treaty signed, and they believe that pushing for deeper cuts at this stage would prove fatal to the wide yet fragile consensus. And, with the example of the Vienna Convention on CFCs in mind, they believe that Kyoto creates a dialogue that can be parlayed into something truly meaningful.

Another concern is the Protocol's carbon budgets for participating countries. These are set relative to 1990 emissions levels, and vary from 92 per cent to 110 per cent. The matter becomes complex when the economies of the nations are taken into account, for the eastern European states have suffered economic ruin since 1990 and are producing 25 per cent less CO_2 than they were then. With their Kyoto carbon budgets set at

8 per cent less than their 1990 levels, they have valuable carbon credits to trade. These credits, which do nothing to diminish climate change, are known as 'hot air'. They constitute a substantial waste in dollars and opportunities to abate emissions. There is another issue here, as many economists argue that the ex-communist countries should not be granted a steady stream of carbon dollars solely on the basis of poor economic performance.

For the first target period of the treaty (2008–2012) the European Union has a carbon budget of 8 per cent less than it emitted in 1990 and the US 7 per cent less. Australia, on the other hand, has a budget of 8 per cent *greater* than it emitted back then (108%). Only Iceland did better than that, with a 10 per cent increase (110%)—though even Norway managed an increase of 1 per cent. Was this a fair outcome, and how did it come about? Some claim that the variations reflect the real costs of compliance for the nations involved, while others see political sleight of hand at work. These complex issues involve details of national economies that are far beyond the scope of this book. In seeking to understand what went on, however, we can examine a single, well-documented instance— the 'special deal' that Australia negotiated for itself—elements of which apply to all the concessions made.

Australia has the highest per capita greenhouse emissions of any industrialised country—25 per cent higher than the US when all sources are accounted for—and Australia's growth in emissions over the last decade has been faster than that of other OECD countries.[118] The Australian delegation to Kyoto argued that this is due to Australia's special circumstances—which include a heavy dependence on fossil fuels, special transport needs (being a large, sparsely populated continent) and an energy-intensive export sector. This all adds up, they stated, to a prohibitively high cost of meeting its Kyoto target, and therefore concessions were needed.

Ninety per cent of Australia's electricity is generated by burning coal. This is a matter of choice rather than necessity, however, for Australia also has 28 per cent of the world's uranium, the world's best geothermal province and a superabundance of high-quality wind and solar resources. Concerns about climate change have been voiced in Australia for over thirty years, and the nation's increased dependence on coal and consequent high cost of shifting to a less carbon-intensive economy were, it now seems, poor economic decisions. Should any country be rewarded for that?

The transport argument is also weak, for while Australia is large, its population is highly urbanised, so 60 per cent of transport fuel is used in urban areas. And as for its energy-intensive exports, Australia is no more exposed here than Germany, Japan or the Netherlands—all strong Kyoto supporters.[118] Coal dependence, transport difficulties and vulnerability in the export sector all boil down to cost, and according to the Australian Bureau of Agricultural and Resource Economics (ABARE) the economic burden Kyoto places on Australia is substantial.

Using the so-called MEGABARE economic model, the bureau predicted that Australia's real gross national expenditure would fall by between one quarter and one half of 1 per cent per annum if a European-style cut in emissions was implemented. This was touted as shocking news by the then Minister for Minerals and Energy, Senator Warwick Parer, who proclaimed in Parliament that this would cost an Australian family of four 'about $7600' per annum—something the electorate would never stand for. Australian National University economist John Quiggin took a closer look at MEGABARE and revealed this to be a distortion. He demonstrated that should Australia's economy grow by an average of 3.5 per cent per year over coming decades, the $7600 would come out of an average family expenditure, in current terms, of $1.86 million![240]

Furthermore, if the nation ratified Kyoto, Quiggin discovered, Australians would have to wait for a doubling of their per capita incomes until 1 March 2025—rather than having it on 1 January of that year—a delay of a paltry two months.[118]

The MEGABARE results brandished during the Kyoto negotiations also conflicted with myriad studies that Australia chose not to promote. Although diverse in their assumptions, these studies showed that Australia could cut its energy consumption and meet its Kyoto target at no net cost whatsoever. As the MEGABARE study came under increasing scrutiny, documents came to light under Australia's Freedom of Information Act revealing how it had been funded, to the tune of $400,000, by the Australian Aluminium Council, Rio Tinto, Mobil and other like-minded groups, all of whom had received a seat on the study's steering committee.[118]

Such was the resistance from the Australian government to Kyoto that Senator Robert Hill (who as Minister for the Environment led the 1997 delegation) knew that only a deal transparently in favour of his nation would stick. No consensus had been reached by the scheduled end of negotiations, and the conference clock was stopped at midnight while delegates argued into the small hours. As the text was read for the final time, Senator Hill rose to his feet and raised a new issue: in Australia's case, land-clearing must be considered. His rationale was that, by protecting forests, Australia was storing CO_2. As land-clearing had declined since the 1990 baseline year, this was akin to eastern European 'hot air', and it would give Australian industry the option of pretty much pursuing business as usual. Faced with either agreeing to this request or seeing the meeting collapse, the delegates agreed to the concession.[118]

Senator Hill was given a standing ovation on his return to Australia, yet his country still refuses to ratify Kyoto, all the while claiming that it will reach its targets anyway! If you are confused by this, don't worry—so is the rest of the world. It's easy to get angry at such a self-interested, muddled approach to negotiations, but we must remember that the outcome may have been nothing more than a fair trade deal. Even so, Australia looks set to reap a bitter harvest, for its stock traders are missing out on an estimated $150 million per year because carbon credits are not being traded on Australian trading floors. Consider too that Japan—which purchases Australian coal—must now buy credits to offset emissions that result from burning that coal, a cost that will doubtless be passed on to Australian coal miners; but because Australia has not ratified Kyoto, no credits can be created there. Instead, the benefit of the credits will flow to a third country—perhaps New Zealand.

So what can be said, by way of summary, of the carbon budgets allocated to the signatory nations under Kyoto? Perhaps they are neither entirely equitable nor fair: but they were agreed to, and as such any debate about their fairness is academic. Only as the treaty matures and commitments for future target dates are set will there be a chance to revisit them.

Another important objection raised by Kyoto's opponents relates to the viability of Kyoto's carbon dollar. A case can be made that developing a new global currency from the top down is too risky to be acceptable. After all, the foundation of any currency is trust—in this case trust that the seller of the carbon credit will do what is required to abate their carbon emissions. What real guarantee do we have that forests will be planted and cared for, or that polluting industrial infrastructure will be dismantled as a result of the sale of carbon credits? Even with goodwill on all sides such schemes may fail because nations such as Russia, where credits may be spent, do not have the legal and regulatory institutions in place to ensure compliance.

Those who endorse the new currency argue that while the risks of creating a carbon dollar are great, the potential benefits are even greater, because carbon trading can dramatically lower the cost of meeting emission targets.[131] And the use of emissions trading as a tool to reduce pollution has a good track record. The system was invented in the US in 1995 to deal with sulphur dioxide pollution from burning coal. It proved enormously successful and it has been adopted for a number of pollutants since.[133] For example, the Chicago Climate Exchange, a voluntary trading scheme involved in the development of the sulphur dioxide markets, traded over 1 million tonnes of CO_2 in its first six months of carbon trading (to 1 July 2004).[235]

This is how emissions trading works: a regulator imposes a permit requirement for the pollutant, and restricts the number of permits available. Permits are then given away on a proportional basis to polluters, or are auctioned off. Emitters who bear a high cost in reducing their pollution will then buy permits from those who can make the transition more easily. Benefits of the system include its transparency and ease of administration, the market-based price signal it sends (which encourages structural adjustment), the opportunities for new jobs and products it creates, and the lowered cost of reducing pollutants.[133]

There are two ways that the new currency could be created: top down or bottom up, and the Kyoto signatories decided on a top-down approach. The difficulties of this become apparent if we look at how similarly ambitious schemes have been implemented.

The European Union, for example, established the euro top down only after developing a strong central bank to handle the currency, and a stringent set of rules which many European nations have difficulty keeping. The General Agreement on Tariffs and Trade (GATT), in contrast, was built from the bottom up, from a series of bilateral deals between trusted partners that were brought under a multilateral umbrella.[119] Some economists argue that this would result in a more stable carbon dollar. They envisage building a carbon currency from a series of deals between trusted partners and, as with the World Trade Organization, new partners could be added as they prove their trustworthiness.

The bottom up approach has much intrinsic merit, but there are two very good reasons why it should not be tried. The first is time. It took fifty years to build GATT, and we don't have anything like that time to create a carbon dollar. The second is that so much effort has already been put into creating a carbon dollar from the top down. To change things now could destroy the only global mechanism in existence to deal with the climate change problem.

A final issue that must be addressed is the breadth of the treaty. The Americans, in the run-up to Kyoto, displayed great anxiety about the exclusion of the developing world from immediate constraints. While it is true that the emissions of many developing nations were not limited, it is only fair to note that 'transitional' nations such as Ukraine, the Czech Republic, Bulgaria and Romania were. The exclusion of the developing world, the Americans claimed, would give them an unfair economic advantage. On 25 July 1997 the US Senate passed a resolution—95 to 0—declaring that it would reject any treaty that did not require 'new specific scheduled commitments to limit or reduce greenhouse gas emissions for Developing Country Parties within the same compliance period'.[119] Republican Senator for Mississippi Trent Lott summed up Senate sentiment when he stated: 'And what would the developing nations contribute? What would our neighbours in Mexico have to do to stop global warming? Nothing. What about other so-called developing nations like Korea, China, India and Brazil? The treaty lets them off the hook.'[119]

These views are important, for they are the stated reason that the world's largest economy refuses to ratify Kyoto, and without US engagement, the treaty's impact on climate change will remain feeble indeed.

Senator Lott's speech appeals to one of humanity's most base instincts—the suspicion of cheating by others. So are the developing nations really 'free riders'? Some experts believe there are valid reasons for excluding the developing nations from the first round. Foremost is the principle of natural justice: the developed world has largely created the problem to date, so it should carry the bulk of the burden. There is also the example of the Montreal Protocol's success with CFCs. Developing nations initially were not bound by it either, yet it proved to be an outstanding accomplishment in tackling the danger presented by the ozone hole.

One of the principal fears surrounding the non-inclusion of the developing nations, in both the US and Australia, is that jobs would go offshore. Of all industries, the most vulnerable to a rise in the cost of electricity is the aluminium smelters.[118] Governments are vigorously lobbied to build more coal-fired power plants to get electricity at rock bottom prices. But even this is not enough. Australian households pay

12–20 cents per kilowatt hour for electricity, while aluminium smelters pay around 2 cents, which means that a significant part of everyone else's power bill is a direct subsidy payment to the smelters.[118] With such unfair distortions in place, it's not clear that exporting such industries would be a bad thing for either the environment or the national economy. Furthermore, it is imperative to get the smelters to pay a fair price for their power, otherwise market forces can never induce them to limit their emissions. Given Kyoto's manifest problems, it may seem best to tax carbon emissions at the smokestack, yet this simple and effective solution finds no favour in Australia or the US.

It is of paramount importance to understand that the Kyoto Protocol is the only international treaty in existence created to combat climate change. For those who urge abandonment or who criticise Kyoto there are two questions: what do you propose to replace Kyoto with, and how do you propose to secure international agreement for your alternative?

TWENTY-FIVE

COST, COST, COST

It is inconceivable that humankind, with all its noble achievements, its aspirations and goodwill, will stay indifferent to the cry of the climate community. The struggle to redress the climate will surely be tackled on several grounds and in such a way as to ensure stability among the climate systems…But most important of all is the fact that we must, imperatively, change our attitudes and agree to live modestly and realistically—all for the sake of the future— which is not ours but which we have borrowed from future generations.

Yadowsun Boodhoo, President of the World Meteorological Organization Commission for Climatology, *World Meteorological Organization Bulletin,* 2003.

The governments of both the US and Australia say they refuse to ratify Kyoto because of prohibitive cost. A strong economy, they believe, offers the best insurance against all future shocks, and both are hesitant to do anything that might slow economic growth.

You might think that this would have precipitated a careful analysis of the costs of ratification versus non-ratification. Nothing of the sort has in fact occurred. Instead, wildly varying estimates have been produced by an array of special interest groups, and it is these that have informed the debate. Consider the estimates produced by William Lash for the Center for the Study of American Business. Lash says ratification would mean a drop in wages growth of 5 to 10 per cent, increased domestic energy costs of 86 per cent, a cut to the average American family income of $2700, a 25 per cent reduction in the domestic consumption of fossil fuels (the equivalent of stopping all road, rail, air and sea traffic permanently) and an increase in farm production costs of $10–20 billion.[211]

The US Department of Energy also sees high costs ahead—around $378 billion annually, yet the Clinton administration costed the excise at around one four-hundredth of that—a billion dollars per year. On the other side there are those—including a coalition of public interest groups—who argue that there may be a positive economic benefit in ratification. They posit that the US could comply and still see domestic energy bills decline by $530 per household per annum.[212]

Some industries also see low costs. Adair Turner, former director general of the Confederation of British Industry, stated:

> If renewable fuels, for instance, cost three times as much as present fossil fuel prices, the impact of Britain switching to a primarily renewable basis by 2050 would be to reduce national income in that year by just 4 percent. This would cut annual growth from now till then by only one-tenth of 1 per cent— implying that we would reach in 2052 the standard of living otherwise attained in 2050.[235]

In Australia, the Allen Consulting Group's 2003 'Sustainable Energy Jobs Report' showed that, with a wise policy mix, including energy-efficiency strategies and sound demand management, building a renewable energy sector can be a net economic positive and create rural jobs.[237] And their study of Victoria's five-star energy rating for housing shows that there are significant benefits for the economy of shifting investment from energy supply to improving efficiency.[238]

With estimates extending from virtual national bankruptcy to an overall benefit, how is the intelligent reader without a degree in economics to discover the truth? There is, thankfully, one certain guide available to all of us—that of past experience.

Economist Eban Goodstein has undertaken a detailed analysis of past projections of regulatory costs as they relate to a variety of industries. Goodstein demonstrated that in every case, when compared with the actual costs paid, the estimates were grossly inflated.[209] His examples range from asbestos to vinyl, and in all instances but one the estimated cost flowing from regulatory change was at least double the actual cost paid, while in some cases estimates were even more exaggerated. This inflation of estimated costs holds regardless of whether industry itself or an independent assessor did the work, which suggests a systematic source of error.

Goodstein argues the reason for this discrepancy is that economists find it difficult to anticipate the innovative ways in which industry goes about complying with new regulations. In some instances they dump the old processes altogether and adopt new, cost-effective ones, while in others they radically transform their entire business. The projections, in contrast, assume a business-as-usual approach that must absorb the burden of costs. Goodstein's analysis of projected versus actual costs for environmental clean-ups provides another interesting outcome. In his study these tasks were almost always underestimated—in some instances grossly so—which leads one to wonder if the economists who calculate the estimates are ignorant of matters of the environment or, more nefariously, have an anti-environment bias.

Experiences such as those documented by Goodstein have prompted other economists, such as Yale University's William Nordhaus and Harvard's Dale Jorgenson, to argue that the emission decreases required to meet the first round of Kyoto targets (to 2012) will be modest. This should assure us that compliance with Kyoto—and indeed deeper emissions cuts—will not bankrupt our nations. It may even, in the longer term, do our economies some good by directing investment into new infrastructure. Yet the cost of compliance is only half of the equation, for to make a truly informed decision about Kyoto—or more radical proposals—we need to know the cost of doing nothing. Neither the US nor Australian government has yet carried out this exercise, though agencies in the US government have been accumulating data that give some indication of what those costs might be.

The National Climatic Data Center lists seventeen weather events that occurred between 1998 and 2002, which cost over a billion dollars apiece. They include droughts, floods, fire seasons, tropical storms, hailstorms, tornadoes, heatwaves, ice-storms and hurricanes; the most expensive, at a cost of $10 billion, was the drought of 2002.[117] This suggests that the costs of doing nothing about climate change are so large that the failure to calculate it bankrupts the argument.

Over the last four decades the insurance industry has been reeling under the burden of losses as a result of natural disasters, of which the impact of the 1998 El Niño offers a fine example. Paul Epstein of the Harvard Medical School calculated that, in the first eleven months of that year alone, weather-related losses totalled $89 billion, while 32,000 people died and 300 million were made homeless. This was more than the total losses experienced in the entire decade of the 1980s.[197]

Since the 1970s insurance losses have risen at an annual rate of around 10 per cent, reaching $100 billion by 1999. Losses at this scale threaten the very fabric of our economic system, for an annual increase in the damages bill of 10 per cent means that the total bill doubles every seven or eight years. Such a rate of increase implies that by 2065 or soon thereafter, the damage bill resulting from climate change may equal the total value of everything that humanity produced in the course of a year.[197]

Illustrative of the rising cost of insurance is the situation of a homeowner in Florida. With extreme weather events on the increase, they now pay a 'deductible' (the amount left to pay in the event of a disaster) on weather-related insurance claims of around $100,000. Both insurance industry and climate trends suggest that householders elsewhere, whose weather-related deductible may now be in the hundreds of dollars, may soon be facing deductibles in the thousands or tens of thousands of dollars. The rising bills result largely from the laws of physics. Consider, for example, the impact of wind speed. An increase in wind speed during a storm from 40–50 knots to 50–60 knots increases building damage by 650 per cent.[130] Similar escalations apply to extreme events as diverse as hurricanes, wildfires, floods and heatwaves. With all projected to increase, the rapid escalation of insurance bills is unavoidable.

Even if costs do not escalate at 10 per cent per annum, the problem will remain substantial. In 2001 Munich Re, the world's largest re-insurance company (re-insurers insure the insurance companies, and thus set insurance rates) estimated that by 2050 the global damage bill from climate change could top $500 billion. Even at these more conservative estimates, insurance industry leaders doubt that their businesses will be able to absorb the claims for much longer.[118]

The re-insurers are fighting back by reviewing their provision of CEOs' professional indemnity insurance based on their efforts to reduce greenhouse gas emissions. Jeffrey Ball wrote in the *Wall Street Journal*, on 7 May 2003:

> With all the talk of potential shareholder lawsuits against industrial emitters of greenhouse gases, the second largest re-insurance firm, Swiss Re, has announced that it is considering denying coverage, starting with directors' and officers' liability policies, to companies it decides aren't doing enough to reduce their output of greenhouse gases.[235]

Christopher Walker, managing director for a unit of Swiss Re, told the *Wall Street Journal*: 'Emissions reductions are going to be required. It's pretty clear.'[235]

Believers in the efficacy of the free markets have suggested that governments should butt out of the regulation of greenhouse gases because market forces will see industries voluntarily reduce their emissions. Despite the efforts of the re-insurers, this view has two serious counts against it. First, in the real world we see little sign of this happening. Secondly, consider how this 'solution' would work if applied to taxation. Why would a voluntary approach to a proposal that adds up to a carbon tax do any better?

With so many analyses demonstrating that rising greenhouse gas emissions are a serious threat to our Earth, and with the cost of reducing carbon dioxide emissions evidently small, you might wonder again why there is such resistance to ratification from the US and Australia. Part of the answer, I think, entails philosophical differences between these countries and Europe.

America and Australia were created on the frontier, and the citizens of both nations hold deep beliefs about the benefits of endless growth and expansion. As a result, both have large immigration programs (Australia's has grown markedly in recent years) and thus high population growth rates relative to their European peers—and this leads to enormous difficulties in adhering to the emission reductions required under Kyoto.

In Australia's case the difference between pursuing an immigration program that would stabilise its population, or see it continue to grow at the rate of 70,000 per annum will, by 2020, increase the nation's emissions by 65 million tonnes of CO_2 per annum. In effect, pursuing population growth is arguably the single biggest impediment to Australia in reaching its Kyoto targets, and thus too it's the primary cause for needing concessions. Put another way, Kyoto questions the philosophies underpinning societies such America and Australia, which cling to the myth of limitless growth.

There is, however, more to humanity's reluctance to tackle climate change than that. If scientists were predicting the imminent return of the ice age I'm certain our response would have been more robust. 'Global warming' creates an illusion of a comfortable, warm future that is deeply appealing, for we are an essentially tropical species that has spread into all corners of our globe, and cold has long been our greatest enemy. From the beginning we have associated it with discomfort, illness and death, while warmth is the essence of everything good—love, comfort and life itself.

Our evolutionary response to the threat of cold is most clearly seen in the young. Children dredged from frozen ponds hours after slipping in have lived because, over millennia, their bodies have evolved defences against the ever-present threat of freezing to death. And, of course, mothers, even in this modern age, do everything they can to protect their offspring from cold. But in today's world this way of thinking can be dangerous, for in industrialised societies sudden infant death syndrome (SIDS) is a far greater threat to the young than freezing is, and in many instances has been attributed to the inadvertent overheating of babies. Laying a child on its back to sleep reduces the risk of SIDS, but the reasons for this, until recently, remained unclear. Physiologists now think that one powerful factor is that heat is more readily lost from the chest and stomach than the back, and with the stomach against the bed, heat transfer is impaired.[199] Sore throats have also been associated with SIDS, and like any bacterial infection they result in a fever, further stressing the infant's heat-shedding mechanisms.

Our deep psychological resistance to thinking that 'warm' might be bad allows us to be deceived about the nature of climate change. Those who have exploited this human blind spot have left many people—even the well-educated—confused. This is the result of an unhealthy, in some instances corrupt, relationship between government and industry. And it is into this cesspit that we must now leap.

TWENTY-SIX

PEOPLE IN GREENHOUSES SHOULDN'T TELL LIES

> The Devil can cite scripture for his purpose.
> An evil soul, producing holy witness,
> Is like a villain with a smiling cheek,
> A goodly rotten apple at the heart
> O, what a goodly outside falsehood hath!
> William Shakespeare, *The Merchant of Venice*

It is in the US, and specifically with the second Bush administration and its industry supporters, that the opposition to reducing emissions of greenhouse gases is most virulent. The American energy sector is full of established, cashed-up businesses that use their influence to combat concern about climate change, to destroy emerging challengers, and to oppose moves towards greater energy efficiency. The fact that in the 1970s the US was a world leader and innovator in energy conservation, photovoltaics and wind technology, yet today is a simple follower, is testimony to their success. It is almost impossible to overestimate the role these industries have played over the past two decades in preventing the world from taking serious action to combat climate change.

The battlefield on which these struggles are taking place is as much the arena of public opinion and closed-door political manoeuvrings as the stock market, and much industry propaganda is very clever. The threat of climate change has been in the public consciousness for several decades now. As early as 1977 the *New York Times* carried the headline 'Scientists Fear Heavy Use of Coal May Bring Adverse Shift in Climate'.[70] But it was not until the late 1980s—when the Montreal Protocol demonstrated that constraints could be applied to damaging emissions and action to restrict greenhouse gases was emerging—that industry embarked upon a propaganda war.

Among the first to move were the US coal producers. Fred Palmer, then head of Western Fuels (now company vice-president at Peabody Energy, the world's largest coal producer) led a campaign—informed apparently by his personal beliefs—that the Earth's atmosphere 'is deficient in carbon dioxide', and that producing more would herald an age of eternal summer. In a move rather like the CEO of an arms manufacturer arguing that a nuclear war would be good for the planet, Western Fuels wanted to lead the charge in creating a world with atmospheric CO_2 of around 1000 parts per million.[81]

Palmer's views were the basis for the propaganda video *The Greening of Planet Earth*, which cost a quarter of a million dollars to make, and which promoted the idea of 'fertilising' the world with CO_2 to boost crop yields by 30 to 60 per cent, thus bringing an end to world hunger. While such ludicrous and bare-faced allegations could be laughed off by scientists, public awareness of the issue was such that many people were misled. *The Greening of Planet Earth* was widely circulated in Washington in the lead up to the 1992 Rio Earth Summit, and among those who saw it were the first President Bush and his chief of staff, John Sununu. Ross

Gelbspan, former editor at the *Boston Globe* and author of a book that unmasks the motives and industry funding of the climate change sceptics, found that the video had a profound impact in Washington. It was, he claimed, Sununu's 'favourite movie', while Bush senior's energy secretary, James Watkins, cited it as a credible source in interviews about climate change.[239]

With the election of George W. Bush, the fossil fuel lobby became even more powerful, and it has been able to corrupt processes within the federal bureaucracy and the soliciting of scientific advice. In June 2005 the *New York Times* ran a story revealing one of the ways that this is done. Philip A. Cooney, a Bush aide and an oil industry lobbyist fighting against the regulation of greenhouse gases, removed or adjusted descriptions of climate research that government scientists and their supervisors (including some senior Bush administration officials) had already approved. Many of Cooney's changes appeared in the final reports, and their overall effect was to minimise concern about climate change.[241] At the most recent count a dozen major reports on climate change have been altered, suppressed or dismissed by the White House, including a ten-year, peer-reviewed study by the IPCC commissioned by Bush senior, studies by the National Academy of Sciences, the National Ocean and Atmospheric Administration and NASA.[110] In September 2002 the White House released the Environment Protection Authority's annual report with the entire section dealing with climate change deleted.

The US government's sympathy for the nonsense touted by Fred Palmer and his ilk is not necessarily a reflection on the intellectual capacities of those involved, but rather its capacity to be bought. Coal miners donated $20 million to the Republican cause in 2000, and have added $21 million since, ensuring that industry access to Vice-President Cheney and his secret energy committee is unparalleled.[110] In 2001 Quin Shea, a lobbyist for the Edison Electric Institute, told a closed-door conference that the Bush administration 'desperately wanted to burn more coal... Coal is our friend', and that to do so they would scuttle Clean Air and

Clean Water Act requirements. In this the administration has been as good as its word for, as Shea quipped, it may be some time before the industry has another president like 'Bush or Attila the Hun'.[110]

Attempts by industry to obtain political influence are not restricted to the US. Australia is the world's largest exporter of coal, and those in the business in this country have also been active. Rio Tinto, the world's largest mining company, owns Australian coal mines and is also a prodigious user of power for smelting. Rio Tinto's chief technologist, Dr Robin Batterham, was appointed Australia's chief scientist by the Howard government, and in this capacity advised on issues such as climate change. More germane, perhaps, to the Australian government's antipathy to reducing emissions, was the 1996 appointment of Senator Warwick Parer as Howard's Minister for Minerals and Energy. Before entering politics Parer was a senior executive with Utah Mining, one of Australia's largest coal producers. He entered the Australian Senate in 1984, but remained chairman of Queensland Coal Mine Management until he became minister. In March 1998 it was revealed he owned $2 million in shares in three Queensland coal mines. This was in breach of the PM's ministerial guidelines, but Parer—who had sacked his press secretary for not declaring shareholdings—was allowed to remain. It was only in October of that year that he quietly resigned as minister, before leaving the Senate in February 2000.[118] His influence was far from finished, however, for he was then appointed by the Howard government to head a review of the energy sector.

The coal industry has not acted alone in misrepresenting the dangers of climate change. Perhaps the greatest damage was done by the Global Climate Coalition, an industry lobby group founded in 1989 by fifty oil, gas, coal, auto and chemical corporations. During the eleven years of its existence the organisation gave $60 million in political donations, and spent millions more on propaganda. The stated purpose of the Global Climate Coalition was to 'cast doubt on the theory of global warming'.[115] It spread misinformation and doubt wherever it could, and among its more effective scare tactics was the claim that addressing climate change could add fifty cents per gallon to the price of gasoline in the US. Its greatest success, however, was the role it played in the 1992 Rio Earth Summit's failure to adopt strong measures to protect all humans from the danger of climate change.

As the scientific evidence for climate change began to firm up, the agenda of the Global Climate Coalition was re-evaluated by some members. DuPont left in 1997: presumably they had learned from their experience with the Montreal Protocol that regulation to control pollutants can be good for business. A few months later BP also broke ranks. Shortly after making his decision, BP CEO Lord Browne of Madingley said, 'We may have left the church in terms of climate change. But it's almost impossible to express the depth of support from within the company for the position we've taken.'[118]

The Global Climate Coalition collapsed in March 2000 when Texaco defected, leaving the group with so few members that it was no longer effective. Among those who stuck it out to the bitter end were Exxon Mobil, Chevron (an oil company) and General Motors. The coalition's website, however, remains active and is as full of misleading material as ever. Visiting it, I am struck by its resemblance to a dinosaur whose brain had been irretrievably damaged, but which still staggers along, wreaking havoc as it wends its way to its grave.

The extent of the division in industry over climate change was revealed in Davos, Switzerland, in early 2000, when world business leaders declared that climate change is the greatest threat facing the world. Later that year a poll of Fortune 500 executives reported that 34 per cent supported ratification of the Kyoto Protocol, with only 26 per cent opposed.[118] Positive industry coalitions are now beginning to form, with seven large American energy and manufacturing corporations establishing the Partnership for Climate Action, which commits them all to reduce their emissions below the national Kyoto target. This is heartening news. Yet reactionary groups have sprung up worldwide to fill the gap left by the implosion of the Global Climate Coalition.

One of the most influential is the Australian-based Lavoisier Group, which was set up in April 2000, and held its first conference a month later under the leadership of former Hawke government finance minister Peter Walsh. The inaugural speech was given by Hugh Morgan, then CEO of Western Mining Corporation, one of Australia's largest mining companies and now part of BHP. Among the Lavoisier Group's many outrageous claims is that the IPCC is an elaborate conspiracy engineered by hundreds of climate scientists who twist their results in order to maintain their research funding. (This singularly spurious claim, incidentally, seems to have struck a chord with Michael Crichton, whose novel *State of Fear* uses it as a major theme.)

Various other groups that contest the issue of climate change include Fred Palmer's Greening Earth Society, Frontiers of Freedom (founded by Republican senator Malcolm Wallop of Wyoming), the Cooler Heads Coalition (responsible for the website globalwarming.org), the Institute of Public Affairs (a right-wing think tank based in Melbourne) and the US Science and Environment Policy Project, long affiliated with Fred Singer, a member of Sun Myung Moon's Unification Church. A brief search for 'climate change' on Google brings up more examples, including 'Myths of Global Warming' by www.biblebelievers.org.au and 'Globaloney Warming' by Liberty Australia.[121] Between them we learn that global warming is a hoax that most scientists don't believe in and that 'The powers that be are using this myth in the Hegelian Dialectic'.[122]

Such gobbledygook is frequently employed to bewilder the general reader, though at times these groups will push that much further. The Leipzig Declaration is a particularly interesting case in point. This document appeared in 1995, penned by Fred Singer, and purported to have the signatures of seventy-nine scientists from leading universities who subscribe to the view that climate change is not a threat. On investigation, however, the majority of the signatories were found not to be scientists, or had not signed the declaration.[118] Scepticism is an indispensable element in scientific inquiry, but when the intention is to mislead rather than clarify, we have not scepticism but deceit.

Some industries that oppose action on climate change use tactics reminiscent of those of asbestos and tobacco companies, who by constantly challenging and clouding the outcomes of research into the link between their products and cancer, succeeded in buying themselves a few more decades of fat profits. Asbestos and cigarettes can kill individuals, but CO_2 emissions threaten our planet. The Fred Palmers of the world have already bought themselves two decades of fat profits, but the cost to the rest of us has been astronomical. Another decade of such profits may cost us the Earth.

We must break now from this catalogue of infamy to examine the workings of the Intergovernmental Panel on Climate Change. The IPCC is not an industry or lobby group. It was established in 1988 and is a joint subsidiary body of the United Nations environment program and the World Meterological Organization. Its workings illustrate how industry uses proxies to slow down, and tone down, the vital work carried out by the group. The Third Assessment Report (TAR) of the IPCC was released in 2001 and is the work of 426 experts, whose conclusions were refereed (twice) by 440 reviewers and overseen by thirty-three editors, before finally being approved by delegates from 100 countries. As you might guess, the report is as dull as dishwater and confines itself to the lowest common denominator.

To understand why this is so, it is important to know the IPCC's *modus operandi* and membership. It comprises scientists, other experts and government representatives, and although industry itself is not directly represented, it gains an effective voice through the government appointees of fossil-fuel dependent nations such as Middle Eastern states and the US. The IPCC's unique structure allows these delegates to exert undue influence, for the organisation works by consensus. I met scientists who were IPCC members at the Hadley Centre in late 2004. They described mind-numbing days devoted to arguing about single, seemingly irrelevant words or sentences. Every word in the organisation's mammoth reports, they asserted, had been debated, with Saudi Arabia, the US and China—the world's largest oil exporter, oil user and burner of coal respectively—eager to water down wording and to slow progress.

Jeremy Legett, who was an eyewitness to the negotiations that led to the 2001 report, says that when pressed for reasons in requesting changes the head of the Saudi Arabian delegation, Mohammed al-Sabban, said, 'Saudi Arabia's oil income amounts to ninety-six per cent of our total exports. Until there is clearer evidence of human involvement in climate change, we will not agree to what amounts to a tax on oil.'[224] Such attitudes are demoralising for dedicated experts, who know that the fate of our planet is at stake. The outcome is that the pronouncements of the IPCC do not represent mainstream science, nor even good science, but lowest common denominator science—and of course even that is delivered at glacial speed.[70] Yet in spite of its faults, the IPCC's assessment reports, which are issued every five years, carry weight with the media and government precisely because they represent a consensus view. If the IPCC says something, you had better believe it—and then allow for the likelihood that things are far worse than it says they are.

What about those industries that are playing an active role in combating climate change? One of the first companies to break from the Global Climate Coalition was BP, whose group chief executive Lord John Browne has long taken a clear-eyed, dispassionate view of climate change. Under his leadership BP has moved 'beyond petroleum', making a 20 per cent cut in its own CO_2 emissions, and a profit in doing so. And BP is now one of the world's largest producers of photovoltaic cells. Lord Browne thinks that 'the reduction of greenhouse gas emissions is a soluble problem, and that it is now time to move beyond the Kyoto debate'.[111]

This commitment to tackling climate change head-on must be viewed in the broader context of British engagement with the issue, and that begins with James Lovelock. It was Lovelock who, as a successful, independent scholar-scientist, thereby commanding respect from those with a free-

market view of the world, convinced Prime Minister Margaret Thatcher
to take the issue seriously. Long before many environmentalists were
even aware of climate change, Mrs Thatcher was advocating a reduction
in CO_2 emissions.

Thatcher's successor Tony Blair, although from the opposite side of
politics, has been even more active. Indeed, of all current political leaders
he has the firmest grasp of the science surrounding the issue. In a recent
address to British industry Blair stated:

> The emission of greenhouse gases…is causing global warming at
> a rate that began as significant, has become alarming and is simply
> unsustainable in the long term. And by long term I do not mean
> centuries ahead. I mean within the lifetime of my children
> certainly; and possibly within my own. And by unsustainable, I do
> not mean a phenomenon causing problems of adjustment. I mean
> a challenge so far-reaching in its impact and irreversible in its
> destructive power, that it alters radically human existence…
> There is no doubt that the time to act is now.[112]

By 2003 Britain's CO_2 emissions had fallen to 14 per cent below what
they were in 1990, bringing the 20 per cent reduction required by 2010
within the nation's grasp. Moreover, the British Royal Commission on
Environmental Pollution's Report on Energy concluded that the United
Kingdom needs to reduce emissions by 60 per cent by 2050, and this is
being taken seriously. As Blair says, 'There are immense business opportunities in sustainable growth and moving to a low carbon economy', a
view borne out by the 36 per cent expansion in national economic growth
during the period emissions dropped 15 per cent.[112] Significant milestones
of this period include the establishment of the Carbon Trust (which helps
business address energy use), an obligation by power suppliers to provide
15.4 per cent of their energy from renewable sources, and significant
investments in developing wave and tidal power.

Britain is also considering expanding its nuclear power capacity, and
with the nation having taken over leadership of the G8 at the beginning

of 2005, it is hoped that further initiatives in combating climate change will be implemented. At Davos in January 2005, Blair signalled to the Bush White House that, if it hoped for further support in the war on terror, it would have to support the war on climate change as well.

Across the globe the majority of industries and governments are taking the middle ground between Bush and Blair, and a large, albeit informal, group of businesses is slowly shifting its stance. Even most fossil fuel industries no longer argue—at least publicly—about the veracity of climate predictions, but instead seek to reassure the public that there is no urgency about the issue. The world has twenty years at least, they say, before change is needed, and by then things will have begun to 'sort themselves out'.

There are even those who argue that the world can have its cake and eat it too, by which they mean that we can burn all of our fossil fuels and still avoid climate change. A great deal hangs on this claim, for if it is falsified then each kilogram of carbon dug from the ground is, given current technology, an irretrievable step towards a hostile geological age—one in which civilisation will struggle to survive. It is now time to examine how industry envisages that our cake might be both eaten and kept, without leaving us holding a plateful of dung.

TWENTY-SEVEN

ENGINEERING SOLUTIONS?

> When you look at all of this, the only reasonable solution is CO_2 sequestration because it is the only one that allows us to continue using fossil fuels and still not harm the climate...It allows using the hydrocarbon resource fully. You can have your cake and eat it, if you want.
>
> Vice-President Dr Philippe Lacour-Gayet (Schlumberger Ltd), Plenary Address, Petroleum Industry Conference, 2004.

By the 1980s the problem of global warming loomed so large that industry, and even some scientists, began casting about for engineering solutions. Here we are talking about true planetary engineering— changing the carbon balance on planet Earth in ways that will affect every living organism—and yet these engineering works were being proposed and tested in the absence of any global body to regulate and approve such activities.

For this reason, and because of an inherent distrust of such solutions, the response of most environmental groups to these initiatives has been lukewarm at best. Yet all would agree we are facing a dire crisis that may require heroic actions to overcome. With many such programs still in the testing or even theoretical stage, the best we can do here is to examine progress to date, and the place to begin is the Southern Ocean.

One of the most ambitious programs proposed to rid the world of excess CO_2 involves fertilising the Southern Ocean with iron filings. The rationale is that iron is the limiting nutrient in sea water, and it's in particularly short supply in the Southern Ocean. Small-scale experiments show that a dusting of iron filings can stimulate spectacular growth in plankton, which captures CO_2 from the surface waters and, when it dies, is carried into the ocean depths. Winds carry iron-rich dust from the world's deserts into the ocean on a huge scale, so the addition of iron filings, proponents argue, is simply an extension of natural processes.

In April 2004 Ken Buesseler of the Woods Hole Oceanographic Institution and his colleagues reported on the results of the Southern Ocean Iron Experiment.[75] Three ships tracked the fate of carbon in a 15-kilometre-square patch of ocean within the Antarctic Circle that had been 'fertilised' with iron filings. This same region had been fertilised on previous occasions and, during the experiment, iron filings were released every four days for seventeen days. Following earlier fertilisations, the plankton had grown well, but there was no evidence that any carbon had moved from the surface layers into the deeper ocean where it might be stored. This is a critical stage in the process, for unless the dead plankton sinks, the carbon it absorbed will simply be re-released to the atmosphere. In the experiment documented by Buesseler, some carbon sank to between 50–100 metres; but was it enough to justify the expense?

As Buesseler and his colleagues put it, 'Using a patch size of 1000 square kilometres…over twenty-one days…resulted in an enhanced flux at 100 metres [depth] of 1800 tonnes of carbon, in response of 1.26 tonnes of iron.'[75] They estimate, however, that only 900 tonnes (roughly half) of that carbon would be sequestered on the sea floor. Given that humans are releasing 13,000,000,000 tonnes (13 gigatonnes) of carbon per year, the disposal of a paltry 900 tonnes through this tedious and expensive process is a poor result indeed. 'It is difficult to see how ocean fertilisation with such a low...export efficiency would...scale up to solve our...global carbon imbalance problem,' the researchers concluded. Even with a more positive result, an unwanted side effect may have been fatal to large-scale implementation: when fertilised, certain kinds of plankton grow at the expense of others, which can lead to an imbalance in the oceans and a loss of biodiversity.

While some researchers have been fertilising the ocean's sunlit layers, others have been pumping compressed CO_2 directly into the ocean depths. This technology was proposed as early as 1977, and for some is the great hope for curing the world's carbon ills. Professor Takashi Ohsumi of Japan's Research Institute of Innovative Technology for the Earth estimates that CO_2 from power plants can be concentrated and liquefied at the cost of around $50 per tonne (though other studies indicate that $100 per tonne is more realistic), and that 'there is no technical barrier envisioned for the implementation of this option either in dissolving CO_2 in the mid-water or placing it on the oceanic floor'.[91] While there may be no 'technical' barriers, preliminary studies indicate that pumping liquefied CO_2 directly into the ocean causes severe side effects.

James Barry of the Monterey Aquarium and his colleagues studied a plume of liquid CO_2 that was released directly onto the sea floor off California at a depth of over three and a half kilometres.[91] They observed 'high rates' of death among organisms in the vicinity of the plume, which seems to have occurred due to the sea water turning acidic (its pH increasing by half to one unit). Barry's team predict high mortality rates for sea creatures wherever the technology is tried out.

Dr Ulf Riebesell of the Leibnitz Institut für Meerewissenschaften in Kiel, Germany, reports that as CO_2 concentrations in the ocean grow, biodiversity will be affected in a number of ways. Species that grow calcite shells, for example, have difficulty surviving in the more acidic water that CO_2 creates. The acid may also harm the long-term growth and reproduction of squid and certain fishes.[93] Despite these early warning signs, Professor Ohsumi appears to think that 'she'll be right, mate', as Australians would say, and that we should go ahead with 'a large scale of CO_2 injection intended to determine ecosystem alteration'.[91]

As the bright possibility of dumping CO_2 into the sea began to dim, the coal industry took up the idea of pumping it underground instead. The process, known as geosequestration, is beguilingly simple in its approach: the industry would simply re-bury the carbon that it had dug up. In truth the history of the technology to date seems impeccable, for oil and gas companies have been pumping CO_2 underground for years, with the Sleipner oilfield in the North Sea the example most frequently cited. The incentive provided by a \$40 per tonne tax on CO_2 emissions by the Norwegian government means that the large volume of CO_2 that comes up with the hydrocarbons is separated off in a highly concentrated form and pumped back into the rocks. At a few other wells around the world (but not at Sleipner) the CO_2 is pumped back into the oil reserve, helping to maintain head pressure, which assists with the recovery of oil and gas, making the entire operation more profitable. Reputedly, 'most' of the CO_2 stays underground. Applying this model to the coal industry, however, is not straightforward.

The problems for coal commence at the smokestack. The stream of CO_2 emitted there is relatively dilute, making CO_2 capture unrealistic. The coal industry has staked its future on a new process known as coal gasification. These power plants resemble chemical works more than conventional coal-fired power plants. In them, water and oxygen are mixed with the coal to create carbon monoxide and hydrogen. The hydrogen is used as a fuel source, while the carbon monoxide is converted to a concentrated stream of CO_2. These plants are not cheap to run: around one quarter of the energy they produce is consumed just in keeping them operating. All indications suggest that building them on a commercial scale will be expensive and that it will take decades to make a significant contribution to power production.

Let's assume that some plants are built and the CO_2 is captured. For every tonne of anthracite burned, around 3.7 tonnes of CO_2 is generated. If this voluminous waste could be pumped back into the ground below the power station it would not matter as much, but the rocks that produce coal are not often useful for storing CO_2, which means that the gas must be transported. In the case of Australia's Hunter Valley coal mines, for example, it needs to be conveyed over Australia's Great Dividing Range and hundreds of kilometres to the west.

Once the CO_2 arrives at its destination it must be compressed into a liquid so it can be injected into the ground—a step that typically consumes 20 per cent of the energy yielded by burning coal in the first place. Then a kilometre-deep hole must be drilled and the CO_2 injected. From that day on, the geological formation must be closely monitored; should the gas ever escape, it has the potential to kill. Miners of old knew concentrated CO_2 as choke-damp, which is an appropriate name, for it instantaneously smothers its victims.

The largest recent disaster caused by CO_2 occurred in 1986, in Cameroon, central Africa. A volcanic crater-lake known as Nyos belched bubbles of CO_2 into the still night air and the gas settled around the lake's shore, where it killed 1800 people and countless thousands of animals, both wild and domesticated. No one is suggesting sequestering CO_2 in volcanic regions such as Nyos, so the CO_2 dumps created by industry are unlikely to cause a similar disaster. Still, Earth's crust is not a purpose-built vessel for holding CO_2, and the storage must last thousands of years so the risk of a leak must be taken seriously.

One of the most disturbing things about this issue is that governments in the US, Australia and elsewhere are debating with industry right now, behind closed doors, on how much risk they will accept on behalf of their constituencies, and how much industry will bear.

Even the volume of CO_2 generated by a sparsely populated country such as Australia beggars belief. Imagine a pile of 200-litre drums, ten kilometres long and five kilometres across, stacked ten drums high. That would be more than 1.3 billion drums, the number required to hold the CO_2 that pours out of Australia's twenty-four coal power stations, which provide power to 20 million people every day. Even when compressed to liquid form, that daily output would take up a cubic kilometre, and Australia accounts for less than 2 per cent of global emissions! Imagine injecting 50 cubic kilometres of liquid CO_2 into the Earth's crust every day of the year for the next century or two.

If geosequestration were to be practised on the scale needed to offset all the emissions from coal, the world would very quickly run out of A-grade reservoirs near power stations and, especially if the power companies are not liable for damages resulting from leaks, pressure would be on to utilise B, C, D and E grade reservoirs. Yet even coal's needs are, on a larger scale, almost negligible, for there are enough fossil fuel reserves on planet Earth to create 5000 billion tonnes of CO_2, a volume of 'cake', to quote Schlumberger's Vice-President Philippe Lacour-Gayet, so prodigious that it seems impossible for Earth to tuck it away without suffering fatal indigestion. All of this suggests that the best case scenario for geosequestration is that it will play a small role (at most perhaps 10 per cent by 2050) in the world's energy future.

Because action is needed now to combat climate change, both the public and the marketplace need to see proof of geosequestration's potential. Big coal should already be building trial coal gasification plants with geosequestration as a test of the economic and technological viability of their approach. Yet, despite offers of government assistance, very little is happening with geosequestration. Typical is Lacour-Gayet's 2004 pronouncement that one of the problems with Kyoto was that it forced us to take action now 'when this is not yet needed'.[174]

Furthermore, Lacour-Gayet's assessment of the cost of geosequestration at just 10 per cent of the value of the energy produced is unbelievable given that 20 per cent of the calorific value of the fuel burned is required merely to compress the CO_2 so it can be injected into the ground.[174] And imagine the cost of building the new generation coal gasification power plants, the separation, storage, pipelines, compressors and injection wells.

Politicians have been seduced by the coal industry's spin. In 2001 Australia's then chief scientist told the Prime Minister's Science Council, in a closed-door session, that geosequestration would increase the cost of generating power from coal by only $5 per megawatt hour. Yet the International Energy Agency was already reporting costs ten to twenty times greater. Following this meeting, the Australian government set up a $500 million research fund for low emission technologies, precisely tailored in its brief to accommodate geosequestration. That's half a billion dollars that will never be fairly shared between all energy options to ensure the best outcome for the nation.

What is at stake in this debate is revealed by a single case study. According to ABARE (August 2004) projections, Australia must increase its power production by more than 50 per cent by 2020 (a slow rate of growth compared with China), and the coal industry would like to secure as large a share of the cake as possible. If they can get the plants built, big coal figures it has at least a half-century's worth of fat profits ahead of it, but opposition is growing. Many people see the building of new coal-fired power plants anywhere as the most damaging thing that can be done to Earth's future. The Sierra Club's Carl Pope says of the situation in the US:

> If approved and built, these [plants] will have operating lifetimes in excess of sixty years. Their carbon dioxide emissions alone will drastically impair the US's ability to cut its emissions. They will also pre-empt the market for wind and solar. So if they are built, we are cooked.[243]

There are other forms of sequestration which are vital for the future of the planet, and which carry no risk. Earth's vegetation and soils are reservoirs for huge volumes of carbon, and are critical elements in the carbon cycle. The development of agriculture has largely been one of using up this resource, and today the world is mostly deforested and its soils exhausted.

Soil carbon can be enhanced by following sustainable agricultural and animal husbandry practices, for this increases the vegetable mould (mostly carbon) in the soil. Lots of carbon—around 1180 gigatonnes—is currently stored this way; more than twice as much as is stored in living vegetation (493 gigatonnes), and storing more seems both simple and desirable.[136] There is real hope for change here, for a vast range of initiatives, from organic market gardening to sustainable rangelands management, are beginning to be seen at the grass-roots level worldwide.

One aspect of this pathway that is being vigorously pursued by some industries is storing carbon in forests and long-lived forest products. This involves either planting forests, or not cutting them down (so that carbon is not released). The Costa Rican government's program to save half a million hectares of tropical rainforest from logging brought it carbon credits equivalent to the amount of CO_2 that would have entered the atmosphere if the forests had been disturbed.[12] And another example is BP's initiative to fund the planting of 25,000 hectares of pine trees in Western Australia to offset emissions from its refinery near Perth.[109] Although forestry plantations are destined to be cut and used, they can be a good short-term store for carbon because the furniture and housing they produce are long-lived, and because the roots of the felled trees (along with their carbon) stay in the ground. The potential to store carbon in roots, however, has been questioned by researchers who discovered that carbon turnover in tree roots is far slower than once imagined, making sequestration by this means less efficient than thought.[140]

There is, however, an overarching concern about storing carbon derived from fossil fuels in forests or in the soil. The carbon in coal has been safely locked away for hundreds of millions of years, and would remain so for millions more had it not been dug up.[172] Yet carbon locked away in forests or the soil is unlikely to remain out of circulation for more than a few centuries. In effect, by trading coal storage for tree storage of carbon, we are exchanging a gilt-edged guarantee for a junk bond.

It's clear that engineering solutions to the carbon problem have proved to be neither as straightforward nor as cost effective as industry would like. Yet scientists continue to work on the problem of safe, secure storage for carbon, and perhaps a solution will eventuate. There is even talk of creating artificial photosynthesis, thereby capturing carbon directly from the atmosphere.

Although such proposals build upon existing technologies, the obstacles are so manifold that implementation on a scale sufficient to combat climate change will surely not arrive before 2050. This suits some industries, for it allows governments to continue throwing billions of tax-payers' dollars into such schemes and, because the mooted solutions are medium to long term, industry can be seen to be doing something and thus retain its social licence to operate. Meanwhile, the competition from less carbon-dense fuels is looking simpler and cheaper by the day.

TWENTY-EIGHT

LAST STEPS ON THE STAIRWAY TO HEAVEN?

> Burning natural gas or oil releases only half as much carbon dioxide (as coal), but unburnt gas is twenty-five times as potent a greenhouse gas as is CO_2. Even a small leakage would neutralise the advantage of gas.
>
> James Lovelock, *Independent,* 24 May 2004.

To people in the petrochemical and motor vehicle businesses, the solution to the climate change problem lies in ascending a metaphorical staircase of fuels which, at each step, contain an ever-diminishing amount of carbon.

Yesterday, the argument goes, it was coal, today it's oil and tomorrow it will be natural gas, with Nirvana being reached when the global economy makes the transition to hydrogen—a fuel that contains no carbon at all.

Although the transition from oil to gas is now well under way, it's been some time coming. For many years the oil companies regarded natural gas as a volatile waste product, to be either burned off or pumped back underground to increase oil pressure at the well-head.[86] Because of its greater hydrogen content, gas burns hotter and cleaner than oil, so it was always valuable stuff; but the technology needed to transport it safely and cheaply did not exist.

One of gas's greatest drawbacks is its low density, which makes it bulky and prone to leaking. It takes a house-sized volume of gas to yield the same energy as a barrel of oil, so barrels—and even tankers—were never an option for its transport. Pipelines were the obvious solution, but suitable pipelines cost around a million dollars for each 1.6 kilometres laid, which meant that until recently, investing a dollar in oil returned twice the profits of one invested in gas.[86]

Technological advances in handling gas, high oil prices, a looming lack of oil and the demand for a cleaner fuel to replace coal have all combined to change the economics of gas, and today it is big business. The most important technical advance involved the refrigeration of gas so that it becomes a supercooled liquid, which permits cost-effective transport, in purpose-built ships, over large distances. With an international trade in shipping developed, and with the larger corporations willing to invest the billions required for gas pipelines, gas appears to be the fuel of choice for the twenty-first century.[86]

Although gas is a more expensive fuel than coal, it has many benefits that make it ideal for producing electricity. Gas-fired power plants cost half as much to build as coal-fired models, and they come in a variety of sizes. Instead of having one massive, distant generator of electricity, as with coal, a series of small gas-fired generators can be dotted about, saving on transmission losses. They can also be fired up and shut down quickly, which makes them ideal for complementing intermittent sources of energy generation such as wind and solar. Furthermore, combined cycle plants, which burn gas to turn a turbine then capture the ultra-hot exhaust emissions to generate more electricity, are extremely efficient at converting fuel to power. If coupled to a heat-using industrial process (called cogeneration) they can achieve efficiencies of 80 per cent. All of this has led Lord Browne, CEO of BP, to comment that 'one dollar invested today in gas-fired generation capacity produces three to four times the amount of electricity [as] the same dollar invested in coal-fired generation capacity'.[86]

Over 90 per cent of new power generation in the US today is gas-fired, and around the world it is fast becoming the favoured fuel. Despite this, gas is not without its problems, including safety issues and the possibility of terrorist attacks on large plants or pipelines. And because methane is a powerful greenhouse gas, its potential to leak must be addressed: parts of the gas infrastructure—such as the old iron pipes used for reticulating gas throughout cities—are decidedly leaky.

Gas is the third step on the stairway to climate-change heaven, but even if all the coal-fired power stations on Earth were replaced with gas-fired ones, global carbon emissions would be cut by only 30 per cent. So despite these savings, if we were to stall on this step of the energy staircase, we would still face massive climate change. In this scenario, a transition to hydrogen is thus imperative; but how likely is it?

In the 1970s the Australian electrochemist John Bockris coined the phrase 'hydrogen economy' and ever since, for many people, hydrogen appears to be the silver-bullet solution to the world's global warming ills. 'Boiled down to its minimalist description,' Bockris wrote, 'the "Hydrogen Economy" means that hydrogen would be used to transport energy from renewables (at nuclear or solar sources) over large distances; and to store it (for supply to cities) in large amounts.'[120] As with so many silver-bullet solutions, however, there's a lot of devil in the detail.

The power source of the hydrogen economy is the hydrogen fuel cell, which is basically a box with no moving parts that takes in hydrogen and oxygen, and puts out water and electricity. While a wondrous-sounding device, it is hardly new technology: the first hydrogen fuel cell, known as a 'gaseous voltaic battery', was built by Sir William Grove in the 1830s. His cell resembled a standard lead-acid battery in that it used sulphuric acid as an electrolyte, but instead of employing lead electrodes it used platinum, which hastens the reaction of hydrogen and oxygen that generates the electricity. The use of such an expensive catalyst was a drawback in developing the technology, but today there are several kinds of fuel cells that use other materials. But whatever their composition, from an economic perspective hydrogen fuel cells can be divided into two types: stationary cells used to produce electricity, and those used in transport.

The most promising cells for the stationary production of electricity are known as molten carbonate fuel cells, which use molten potassium carbonate instead of sulphuric acid, and nickel in place of platinum. They operate at a temperature of around 650°C, and although highly efficient (possessing an electric efficiency of around 50 per cent) they take a long time to reach working temperature. They are also very large—a 250-kilowatt model is the size of a railway carriage—making them unsuitable for use in motor vehicles.

Several demonstration projects based on this technology already exist, and a commercial stationary hydrogen cell (using an earlier technology) has been in operation in the US since 1999. It is predicted that a decrease in cost resulting from economies of scale will soon lead to more widespread use of the cells.[120] Although this represents a tremendous technological advance, it does nothing immediate to abate CO_2 emissions, for the hydrogen used today comes from re-forming natural gas. Because some of the energy in the gas is used in this process, and all of the CO_2 it produces is released into the atmosphere, from a climate perspective the world would be better off burning the gas directly to create electricity.

But let's consider hydrogen as a transport fuel. A number of motor vehicle manufacturers, including Ford and BMW, are planning to introduce hydrogen-fuelled, internal combustion engine cars to the marketplace; and the Bush administration plans to invest $1.7 billion to build the hydrogen-powered FreedomCAR. Even so, the use of hydrogen as a transport fuel is at a far more rudimentary stage of development than the technology using stationary cells.[120]

The fuel cell type best suited for transport purposes is known as a proton exchange membrane fuel cell. It is much smaller than the molten carbonate cell and operates at around 66°C, thus being ready for action soon after turning the ignition. These cells, however, require very pure hydrogen. In current prototypes this is supplied from a built-in 'reformer' that converts natural gas or petrol to hydrogen, which again means that, from a climate perspective, we would be better off burning these fuels directly to drive the engine. The best energy efficiency obtained by proton exchange membrane fuel cells is 35 to 40 per cent—about the same as a standard internal combustion engine.

Vehicle manufacturers hope to do away with the on-board reformer required by the prototypes and envisage fuelling the vehicles from hydrogen 'pumps' at fuel stations. There are several ways that this could be done. The one most closely resembling the current system of fuelling vehicles involves producing the hydrogen at a remote central point and distributing it to fuelling stations; and it's here that the difficulties involved in moving such low-density fuel become evident.

The ideal way to transport it is in tanker-trucks carrying liquefied hydrogen but, because liquefication occurs at −253°C, refrigerating the gas sufficiently to achieve this is an economic nightmare. Using hydrogen energy to liquefy a kilogram of hydrogen consumes 40 per cent of the value of the fuel. Using the US power grid to do so takes 12–15 kilowatt hours of electricity, and this would release almost ten kilograms of CO_2 into the atmosphere. Around 3.5 litres of petrol holds the equivalent energy of one kilogram of hydrogen. Burning it releases around the same amount of CO_2 as using the grid to liquefy the hydrogen, so the climate change consequences of using liquefied hydrogen are as bad as driving a standard car.

One solution may be to pressurise the hydrogen only partially, which reduces the fuel value consumed to 15 per cent, and the canisters used for transport can be less specialised. But even using improved, high-pressure canisters, a 40-tonne (40,000 kilogram) truck could deliver only 400 kilograms of compressed hydrogen, meaning that it would take fifteen such trucks to deliver the same fuel energy value as is now delivered by a 26-tonne petrol tanker. And if these 40 tonne trucks carried the hydrogen 500 kilometres, the energy cost of the transport would consume around 40 per cent of the fuel carried.

Further problems arise when you store the fuel in your car. A special fuel tank carrying hydrogen at 5000 psi (near the current upper limit for pressurised vessels) would need to be constructed and be ten times the size of a petrol tank. Even with the best tanks, around 4 per cent of fuel is likely to be lost to boil-off every day. A good example of the rate of evaporative loss of hydrogen occurs whenever NASA fuels the space shuttle. Its main tank takes 100,000 litres of hydrogen, but an extra 45,000 litres must be delivered at each re-fuelling just to account for the evaporation rate.[120]

Pipelines are another option for transporting hydrogen, but as with gas they are expensive—they must be large and built from materials resistant to hydrogen (it makes steel, for example, very brittle). They must also be of high integrity, because hydrogen leaks so easily. Even if the pre-existing gas pipeline network could be reconfigured to transport hydrogen, the cost of providing a network running from central producing units to the world's fuelling stations would be astronomical.

Perhaps hydrogen could be produced from natural gas at the petrol station. This would do away with the difficulties of transporting it, but this process would produce 50 per cent more CO_2 than using the gas to fuel the vehicle in the first place. Hydrogen could also theoretically be generated at home using power from the electricity grid, but the price of electricity for domestic use, and the high cost of hydrogen generation and purification units, would make it prohibitively expensive. Furthermore, the electricity in the grid in places such as the US is largely derived from burning fossil fuels, so home generation of hydrogen under current circumstances would result in a massive increase in CO_2 emissions.

And there is another danger with home-brewing hydrogen. The gas is odourless, leak-prone, highly combustible and it burns with an invisible flame. Firemen are trained to use straw brooms to detect a hydrogen fire; when the straw bursts into flames you have found your conflagration.

Let's imagine for a moment, however, that all of the delivery problems relating to hydrogen are overcome, and you find yourself at the wheel of your new hydrogen-powered four-wheel-drive. Your fuel tank is large and spherical, because at room temperature hydrogen takes up around three thousand times as much space as petrol. Now consider that a call on your mobile phone, the static electricity generated by sliding over a car seat, or even an electrical storm 1.6 kilometres away, all carry a sufficient charge to ignite your hydrogen fuel. When viewed in this light, the thought of a hydrogen car accident hardly bears thinking about. Even garaging your new vehicle means trouble. Current codes for hydrogen storage in the US are onerous, requiring—among other things—expensive ventilation and explosion-proof equipment.[120] This means that unless codes are relaxed a plethora of infrastructure from garages to road tunnels will require modification.

Even if hydrogen is made safe to use, we are still left with a colossal CO_2 pollution issue, which was exactly the opposite of what we set out to do. The only way that the hydrogen economy can help combat climate change is if the electricity grid is powered entirely from carbon-free sources. And this means acceptance of and investment in a series of technologies ranging from solar to nuclear. Strangely, neither the US government nor the vehicle manufacturers have shown much interest in laying the groundwork for this essential prerequisite for transition to the hydrogen economy.

5

THE SOLUTION

TWENTY-NINE

BRIGHT AS SUNLIGHT, LIGHT AS WIND

Once we open the door to consider catastrophic changes, a whole new debate is engaged. If we do not know how human activities will affect the thin layer of life-supporting activities that gave birth to and nurture human civilisation and if we cannot reliably judge how potential geophysical changes will affect civilisation and the world around us…should we not be ultra-conservative and tilt towards preserving the natural world at the expense of economic growth and development? Do we dare put human betterment before the preservation of natural systems and trust that human ingenuity will bail us out should Nature deal us a nasty hand?

William Nordhaus, *Climate Change*, 1996.

One of the key decisions in our war on climate change is whether to focus our efforts on transport or the electricity grid. Many would argue we must do both, and I would agree, if we had the resources and the time. But when it comes to the really big effort required to stop carbon emissions from one or the other, decarbonising the power grid wins hands down. For with that achieved, we can use the renewable power thus generated to decarbonise transport.

Researchers Steven Pacala and Robert Socolow from Princeton University investigated whether the world possesses the technologies required to run an electricity network of the extent, scale and reliability of that we currently enjoy, while at the same time making deep cuts in CO_2 emissions. They identified fifteen basic kinds of technologies, ranging from sequestration to wind, solar and nuclear power, which could play a vital role. Not all of these technologies need to be used, but at least half of them do if we wish to control the world's carbon emissions for at least the next fifty years.[161] 'It certainly explodes the idea that we need to do research for a long time before getting started,' was how Socolow summarised his work. The many examples of governments and corporations around the world that have slashed emissions (by over 70 per cent in the case of some British local councils) while at the same time experiencing strong economic growth show that Socolow is right: big coal and big oil's scare campaign that it's all too hard and too expensive is swiftly being unmasked.

The technologies fall into two lots: those that at present provide power intermittently; and those that can provide a continuous output of power regardless of circumstance. Of all the sources of intermittent power, the most mature and economically competitive is wind, and nowhere has it been pursued with such vigour as in Denmark, the home of the modern wind industry.

At the time the Danes decided to back wind power, the cost of electricity produced this way was many times greater than that produced by fossil fuels. The Danish government, however, could see its potential and supported the industry until costs came down. Today Denmark leads the world in both wind power production and the building of turbines; and wind now supplies 21 per cent of the country's electricity. One striking aspect of the way that wind power has developed there is that some 85 per cent of capacity is owned by individuals or wind co-operatives, and so power lies literally in the hands of the people.

In several countries wind power is already cheaper than electricity generated from fossil fuel, which helps account for the industry's phenomenal growth rate of 22 per cent per annum.[126] It has been estimated that wind power could provide 20 per cent of the energy needs of the United States, and such are their economics that the Clinton government's aim of producing 5 per cent of the nation's electricity from wind by 2020 may yet be realised. Over the next few years the unit price of wind energy is expected to drop a further 20–30 per cent, which will make it even more cost effective.

Wind power, however, is widely perceived as having a major disadvantage—the wind doesn't always blow, which means that it is unreliable. This is a gloss of a more complex reality, for while wind does not blow at the same location with consistent strength, if you take a regional approach it is fairly certain that the wind will be blowing somewhere. Thus, the more widely dispersed wind turbines are, the more they come to resemble base-load providers like coal. One implication of this is that there is lots of redundancy in wind generation, for often there will be several turbines lying idle for each one working at capacity.

In the UK the average turbine generates at only 28 per cent of its capacity over the course of a year. To determine how significant is this disadvantage, it is worth remembering that all forms of power generation have some degree of redundancy. In the UK nuclear power works at around 76 per cent, gas turbines 60 per cent, and coal 50 per cent of the time. The high redundancy in wind, however, is somewhat offset by the high reliability: wind turbines break down far less frequently, and they are cheaper to maintain than coal-fired power plants.[132] One proposal to decrease redundancy is to use excess wind power to pump compressed air into the ground where it could be called on later to drive generators. Another is to create hydrogen, which can then be used to drive stationary power cells at times of low wind activity.

Wind power, unfortunately, has been beset with bad press, including allegations that wind turbines kill birds, and are noisy and unsightly. The truth is, any tall structure represents a potential hazard to birds, and early wind towers did increase that risk. They had a latticework design, allowing birds to nest in them, but they have now been replaced by smooth-sided models. Moreover, all risks need to be measured against each other. Cats kill far more birds in the US than do wind farms.[132] And if we continue to burn coal, how many birds will die as a consequence of climate change? As to noise pollution, you can have a conversation at the base of a tower without having to raise your voice, and new models reduce the sound even further. And in terms of their alleged unsightliness, beauty is surely in the eye of the beholder. What is more unsightly—a wind farm or a coal mine and power plant? Besides, none of these issues should be allowed to decide the fate of our planet.

From the wind let's turn to three important technologies that directly exploit the Sun's power. These are solar hot water systems, solar thermal devices and photovoltaic cells. Solar hot water is the simplest and, in many circumstances, the most cost-effective method of using the Sun's power for household purposes: making it the best way to make large, easy savings in most household power bills. Solar hot water systems sit on a north-facing (in the Northern Hemisphere south-facing) roof, and trap the Sun's rays that are then used to heat water. They require no maintenance, and to ensure that hot water is available whenever you need it, they include a gas or electric booster.

Solar-thermal power stations produce large amounts of electricity— far more than one household could ever use—and they work by concentrating the Sun's rays onto small, highly efficient solar collectors. Their name comes from the fact that they produce both electricity and heat, the heat often being used for some ancillary purpose such as treating effluent. There are many designs in the marketplace at present, and they are rapidly becoming more affordable. In future, solar thermal power plants can be expected to compete with wind for a slice of the grid, and wind and solar thermal are perfect partners in that regard, for if the wind isn't blowing, there's a good chance that the Sun will be shining.

Finally there is the technology that most people recognise as true 'solar' power—photovoltaic cells. Generating your own electricity with photovoltaics is a bit like home brewing, in that once you've bought your equipment you can thumb your nose at the multinationals. It is also simple and (unless you are not connected to the grid and need a battery bank) maintenance-free, while the panels carry a twenty-five-year guarantee, and are likely to last up to forty years.

There are several types of photovoltaic cell on the market, but all act by using the sunlight that falls on them to generate electricity. That electricity must then be transformed into an alternating current of the correct voltage for your area using an inverter. If you are on the grid, all you need are these two items and a power socket, and you can generate power. The average home requires around 1.4 kilowatts (1400 watts) of power to run, and the average size of panels is 80 or 160 watts. Ten of the larger size should do the job, though it is amazing how much more power conscious you become (and thus how much power you save) when you are generating your own.

Photovoltaics operate best in summer, when that extra power for air conditioning is needed. This allows the owner of a photovoltaic system to make money: in Japan you can sell excess power to the grid for as much as $50 per month, and similar schemes exist in fifteen other countries.[126] In 2003, in northern countries, solar power was around eight times more expensive than conventional power, and in Australia, it is four times as expensive. But the cost of photovoltaics is falling so rapidly that electricity generated by this means is expected to be cost effective as early as 2010.

There are, of course, many kinds of power generation not discussed here, including solar chimneys and tidal and wave power, and in certain locations all of these options are now, or soon will be, producing renewable power. If the renewables sector offers one lesson, it is that there is no silver bullet for decarbonising the grid: rather we will see a multiplicity of technologies used wherever favourable conditions prevail.

THIRTY

NUCLEAR LAZARUS?

> We hear the Secretary of State [John Foster Dulles] boasting of his brinksmanship—the art of bringing us to the edge of the nuclear abyss.
>
> Adlai Stevenson, *New York Times*,
> 26 February 1956.

It's often said that the Sun is nuclear energy at a safe distance. In this era of climate crisis, however, the role of Earth-based nuclear power is being reassessed, and what was until recently a dying technology may yet create its own day in the sun.

The revival began in earnest in May 2004, when environmental organisations around the world were shocked to hear the originator of the Gaia hypothesis, James Lovelock, deliver a heartfelt plea for a massive expansion in the world's nuclear energy programs. Lovelock did so, he said, because he believed that climate change was advancing so rapidly that nuclear power was the only option available to stop it. He compared our present situation with that of the world in 1938—on the brink of war and nobody knowing what to do. Organisations such as Greenpeace and Friends of the Earth immediately rejected his call.

Yet Lovelock has a point, for all power grids need reliable 'baseload' generation, and there remains a big question mark over the capacity of renewable technologies to provide it. France supplies nearly 80 per cent of its power from nuclear sources, while Sweden provides half and the UK one quarter. Nuclear power already provides 18 per cent of the world's electricity, with no CO_2 emissions. Its proponents argue that it could supply far more, but even the Bush administration's energy forecasters believe that its share will in fact fall—to just 10 per cent of production—within the next decade.[86]

In discussing nuclear power as a means of creating electricity, we must keep in mind that nuclear power plants are nothing more than complicated and potentially hazardous machines for boiling water, which creates the steam used to drive turbines.

As with coal, nuclear power stations are very large—around 1700 megawatts—and with a starting price of around US $2 billion apiece they are expensive to build. The power they generate, however, is at present competitive with that generated from wind. Because they are large, and many factors relating to safety must be considered, the permitting process for a nuclear power station can take up to a decade, with construction taking around five years. With a fifteen-year gestation period before any power is generated, and even longer before any return on the investment is seen, nuclear power is not for the impatient investor. It is this, as much as concerns about safety, which explains why no new reactors have been built for twenty years in either the US or UK.

Three factors loom large in the mind of the public, however, whenever nuclear power is mentioned—safety, disposal of waste and bombs. The horror of the 1986 Chernobyl disaster in Ukraine was a catastrophe of stupendous proportions whose consequences, two decades after the accident, just keep growing. Thyroid cancer is a rare illness, with just one in a million children developing it spontaneously. But one third of children under four years old who were exposed to fallout from Chernobyl will develop the disease. Seven per cent (some 3.3 million people) of the population of Ukraine have suffered illness as a result of the meltdown, while in neighbouring Belarus, which received 70 per cent of the fallout, the situation is even worse. Only 1 per cent of the country is *free* from contamination, 25 per cent of its farmland has been put permanently out of production, and nearly one thousand children die each year from thyroid cancer. Currently, 25 per cent of the Belarus budget is spent on alleviating the effects of the disaster.[62]

In the US and Europe, safer reactor types predominate but, as the Three Mile Island incident shows, no one is immune to accident, or to sabotage. With several nuclear reactors in the US located near large cities, there are real concerns for a possible terrorist attack.[110] In summarising the situation for nuclear power as it stood in late 2004, the US National Commission on Energy Policy said, 'One would want the probability of a major release of radioactivity, measured per reactor per year, to fall a further tenfold or more [before considering a doubling or tripling of nuclear power capacity]. This means improved defences against terrorist attack as well as against malfunction or human error.'[226]

The management of radioactive waste is another issue of concern. The nuclear industry in the US long looked to the proposed high-level radio-active waste-dump at Yucca Mountain, Nevada, as a solution. But the waste stream has now reached such proportions that even if Yucca Mountain were opened tomorrow it would be filled at once and another dump would be needed. In reality the opening of the Yucca Mountain dump looks to be delayed for years as challenges drag on through the courts. And the problem of what to do with old and obsolete nuclear power plants is almost as intractable: the US has 103 nuclear plants that were originally licensed to operate for thirty years, but are now slated to grind on for double that time. This ageing fleet must be giving the industry headaches, especially as no reactor has ever yet been successfully dismantled, perhaps because the cost is estimated to be around $500 million a pop.

The majority of new nuclear power plants are being built in the developing world, where a less tight-laced bureaucracy and greater central control makes things easier. China will commission two new nuclear power stations per year for the next twenty years, which from a global perspective is highly desirable, for 80 per cent of China's power now comes from coal. India, Russia, Japan and Canada also have reactors under construction, while approvals are in place for thirty-seven more in Brazil, Iran, India, Pakistan, South Korea, Finland and Japan. Providing the uranium necessary to fuel these reactors will be a challenge, for world uranium reserves are not large, and at the moment around a quarter of the world's demand is being met by reprocessing redundant nuclear weapons. This brings us to the issue of nuclear weapons getting into the wrong hands. As the current dispute over the proposed Iranian reactor indicates, anyone who possesses enriched uranium has the potential to create a bomb. As reactors proliferate and alliances shift, there is an increasing likelihood that such weapons will be available to those who want them.

The nuclear industry hopes that technological developments will lead to foolproof reactor types that produce electricity at a cost equivalent to coal. New reactor types include pebble bed reactors, which utilise low-enriched uranium and can be built on a smaller scale than conventional plants, and pressurised water reactors, one of which will soon be built in Normandy, France, which promises to produce power more cheaply than coal. As with geosequestration, however, these are technologies for the future.

What role might nuclear power play in averting the climate change disaster? China and India are likely to pursue the nuclear option with vigour, for there is currently no inexpensive, large-scale alternative available to them. Both nations already have nuclear weapons programs, so the relative risk of proliferation is not great. In the developed world, though, any major expansion of nuclear power will depend upon the viability of new, safer reactor types.

There is one other option for the continuous production of power. Geothermal energy has a long history, yet despite the considerable amount of heat lying between our feet and our planet's molten mantle, geothermal technologies provide a mere 10,000 megawatts of power worldwide.

This sorry state of affairs may soon change, for it now transpires that we have been looking for heat in the wrong places. Previously geothermal power has come from volcanic regions, where aquifers flowing through the hot rocks provide superheated water and steam. It seems sensible to seek geothermal power in such places, but consider the geology. Lava volcanoes only exist where the Earth's crust is being torn apart, allowing the magma below to come to the surface. Iceland, which formed from the ocean floor where Europe and North America are drifting away from each other, is an excellent example of this. There is plenty of heat in such places, but also formidable impediments to generating power, with the biggest problem being the aquifers. Although many flow freely when first tapped, they quickly dwindle, leaving the power plant without a means of transferring the rock's heat to its generators. In the 1980s operators began to pump water back into the ground in the hope that it would be reheated and could be reused. Quite often the water just vanished, for in regions where the Earth's crust is being torn apart many vertical faults exist, and the water was diverted by these rather than returned to the well-head.

In Switzerland and Australia, companies are finding commercially useable heat in the most unlikely places. When oil and gas companies prospected in the deserts of northern South Australia, nearly four kilometres below the surface, they discovered a body of granite heated to around 250°C—the hottest near-surface, non-volcanic rock ever discovered.

The heat had been generated by the natural radioactivity of the granite, which had been kept in place by a blanket of sediment nearly four kilometres thick. What really excited the geologists was that the granite was not in a region where the Earth's crust was being torn apart, but where it was being compressed. This led to horizontal, rather than vertical fracturing of the rock. Even better, the rocks are bathed in super-heated water under great pressure, and the horizontal fracturing meant that it could be readily recycled.

This one rock body in South Australia is estimated to contain enough heat to supply all of Australia's power needs for seventy-five years, at a cost equivalent to that of brown coal, without the CO_2 emissions. So vast is the resource that distance to market is no object, for power can be pumped down the power line in such volume as to overcome any transmission losses.

With trial power plants scheduled for construction in 2005, the enormous potential of geothermal power is about to be tested. Geologists around the world are scrambling to prospect for similar deposits, as the extent of the resource is hardly known. There is some reason to believe, however, that Australia may be specially blessed with this type of potential power, for the continent has been moving northwards at around eight centimetres per year for the past 40 million years, and when it bumped into Asia 15 million years ago, enormous compressional forces were generated. As a result, in Australian mines one kilometre deep, engineers must deal with compressional forces encountered five kilometres down in places such as South Africa.

While this appears to be an exciting breakthrough, we must remember that so far very little electricity has been provided by this form of geothermal heat, and even if successful, it will in all likelihood be decades before this technology is contributing significantly to the world grid.

The power technologies I have discussed place humanity at a great crossroads. Trillions of dollars will need to be invested to make the transition to the carbon-free economy and, once a certain path of investment is embarked upon, it will gather such momentum that it will be difficult to change direction.

So what might life be like if we choose one over the other? In the hydrogen and nuclear economies the production of power is likely to be centralised, which would mean the survival of the big power corporations. Pursuing wind and solar technologies, on the other hand, opens the possibility that people will generate most of their own power, transport fuel and even water (by condensing it from the air).

If we follow this second path, we will have opened a door to a world the likes of which have not been seen since the days of James Watt, when a single fuel powered transport, industrial and domestic needs alike, the big difference being that the fuel will be generated not by large corporations, but by every one of us.

THIRTY-ONE

OF HYBRIDS, MINICATS AND CONTRAILS

> What is it that roareth thus?
> Can it be a Motor Bus?
> Yes, the smell and hideous hum
> Indicat Motorem Bum…
> How shall wretches live like us
> Cincti Bis Motoribus?
> Domine, defende nos
> Contra hos motores bos!
> > A. D. Godley, 'The Motor Bus'

So how do we go about decarbonising our transport systems? Because some forms of transport, such as air travel, depend on high-density fuels (fuels that pack a lot of punch for their volume) this is a thorny question. Some attempts to answer it involve brewing up tailormade fuels from biomass or other renewable sources, and coal miners are also investigating the possibility of tailormaking transport fuels from coal.

Among those pursuing renewables, the Brazilians have the lead, for their vehicle fleet runs largely on ethanol derived from sugar cane—which grows in Brazil better than just about anywhere else. In the US ethanol is largely derived from corn, but the amount of fossil fuel put into growing the crop means that the use of corn-derived ethanol in transportation provides little in the way of carbon savings. If a highly efficient source of ethanol—perhaps switchgrass—can be cultivated, the crop would have to make up 20 per cent of all productivity on land to power the world's cars, ships and aircraft. Humans are already consuming more of the planet's resources than is sustainable, so providing this extra biological productivity will be hard indeed.

Despite such problems, technological advances in transport are so rapid that ways forward can be glimpsed, and nowhere is that so clear as in the Japanese automotive sector.

While companies like Ford have been investing in hydrogen and teams of lawyers to fight higher mileage standards, Toyota and Honda have been hiring engineers to design more efficient cars. As a consequence, they have brought a revolutionary new technology to market which halves fuel consumption and opens the way to astonishing future developments.[118] Known as hybrid fuel vehicles, these new automobiles pair a petrol-driven engine with a revolutionary electric motor.

Driving the Toyota Prius is unnerving at first, for there is no roaring or hideous hum. Instead, when slowing or stopped in traffic, the 1.5 litre petrol engine shuts down and doesn't commence operation again until speed has been built up. The silent electric motor takes over, which is powered by energy generated in part from braking—energy wasted in an ordinary vehicle. The Prius has taken the market by storm and, with a tank that needs refuelling every thousand kilometres, it's the least carbon-costly automobile of its size available, or likely to be available for the next several decades. Relative to the Toyota Landcruiser (or other four-wheel-drives so popular in the US and Australia today) the Prius cuts fuel use and CO_2 emissions by around 70 per cent. That is the same as the amount scientists consider is required for the world economy by

2050 in order to stabilise climate change. If you wish to make a real contribution to combating climate change, don't wait for the hydrogen economy—buy a hybrid fuel car.

If the grid were to be decarbonised, many other transport options would become attractive. Electric cars have been on the market for years, and France already has a fleet of 10,000 such vehicles. But even more exciting technologies are coming out of Europe, including the experimental compressed-air car being developed by Luxembourg-based automobile maker Moteur Developpment International.

These vehicles use the same technology for their tanks that methane-fuelled buses now use to carry compressed gas. The first compressed-air models are a three-seater known as the MiniCATs, which will sell for around US $10,000, and the six-seat CitiCATs, which will retail for around US $16,000. CATS stands for Compressed Air Technology System, and both models are scheduled for sale in France soon, and also to have hybrid petrol engines, thereby extending range and performance.

With a top speed of around 120 kilometres per hour they are not sluggish, and existing technologies can deliver ranges of around 300 kilometres at 50 kilometres per hour, while the cost of refuelling is a mere $2.50. Refuelling with a commercial compressor takes three minutes, and around three and a half hours for a home-based model. This, remember, is the equivalent of the T-model Ford for compressed-air travel, and there will likely be dramatic improvements in years to come. Of course, with no combustion the engine oil doesn't need to be changed in under 50,000 kilometres, while all that comes out of the tail-pipe is pure, cold air.

Imagine what the CitiCAT might mean for a family living in Denmark. They may well own a share in a wind generator, which is used to power their home, and could be used to compress air for their transport fuel as well. Contrast this with the average American family, who, even if nuclear and hydrogen options come to fruition, will continue to purchase their electricity and transport fuels from large corporations. By combating climate change we can not only save our planet, but open the way for a very different future as well.

What about other growing transport sectors such as shipping and air travel? One of the foulest pollutants on Earth is the fuel oil that powers shipping. Over the past few years the volume of international shipping has grown by 50 per cent, meaning that cargo ships have become a leading source of air pollution.[62] The matter that powers these vessels is the left-overs from the production of other fuels, and so thick and full of contaminants is it that it must be heated before it will even flow through a ship's pipes. Satellite surveillance reveals that many of the world's shipping lanes are blanketed in semi-permanent clouds that result from the particulate emissions from ships' smokestacks. Yet solving this problem is potentially easy; after all, until little more than a century ago maritime transport was wind-powered. Using modern wind and solar technologies and energy-efficient engines, sea cargo may, by the middle of this century, once again be travelling carbon-free.

Air travel requires large amounts of high-density fuel of a type that at present only fossil fuels provide. It is also increasing in volume every year. In 1992 air travel was the source of 2 per cent of CO_2 emissions. And in the US, where air traffic already accounts for 10 per cent of fuel use, the number of passengers transported is expected to double between 1997 and 2017, making air transport the fastest growing source of CO_2 and nitrous oxide emissions in the country.[62] Across the Atlantic, by 2030, a quarter of the UK's CO_2 emissions may come from air travel.[112]

The cocktail of chemicals that comprise aircraft emissions work in somewhat opposite ways. Because most modern jets cruise near the tropo-sphere, the water vapour, nitrous oxide and sulphur dioxides they emit have particular impacts. The nitrous oxide emitted by aircraft may enhance ozone in the troposphere and lower stratosphere, yet deplete it further in the upper stratosphere; sulphur dioxide will have a cooling effect.

What is emerging as a most important emission, is water vapour, which can be observed as aircraft contrails.[82] Under certain conditions contrails give rise to cirrus clouds. These clouds cover around 30 per cent of the planet, and while the extent that aircraft contribute to cirrus cloud cover is uncertain, it could be as much as 1 per cent which, because it is concentrated in the Northern Hemisphere mid-latitudes, may have a significant impact on climate.[82] If aircraft were to fly lower, cirrus cloud formation could be cut in half and CO_2 emissions lowered by 4 per cent, while average flight times over Europe would vary by less than a minute.[48]

As mentioned earlier, the potential of these clouds to affect climate was demonstrated in 2001 between September 11 and 14, when the US air fleet was grounded. Average daytime temperatures rose abruptly by 1°C, in such a manner as could not be explained by other factors. This suggests that contrails may be masking the impacts of warming caused by CO_2. Perhaps we will need to maintain them while we reduce our carbon intensity. Equally, there seems no way at present to get aircraft to run on a less damaging substitute for fossil fuel. Without a return to the more leisurely days of travel by zeppelin, air travel will remain a source of CO_2 emissions long after other sectors have transformed to a carbon-free economy.

Transport accounts for around a third of global CO_2 emissions. Transportation by land and sea can easily be powered in ways that emit less CO_2, and the technologies to achieve this either already exist or are on the horizon. Air transport, however, is fast growing and not likely to be fuelled by anything but fossil fuels. Thankfully, jet contrails contribute to global dimming, so it may be just as well that the jets keep flying long after wind-powered and solar-powered ships and compressed-air cars monopolise surface transport.

THIRTY-TWO

THE LAST ACT OF GOD?

> An Act of God was defined as *something which no reasonable man could have expected.*
> A. P. Herbert, *Uncommon Law*, 1935.

Some time this century the day will arrive when the human influence on the climate will overwhelm all natural factors. Then, the insurance industry and the courts will no longer be able to talk of Acts of God, because even the most unreasonable of us could have foreseen the consequences. Instead, the judiciary will be faced with apportioning guilt and responsibility for human actions resulting from the new climate. And that, I think, will change everything.

Pretend, for a moment, that you are a camel-herder living in the Sudan. For all of your life you have known nothing but bad seasons, and in desperation you turned your herds onto the lands of the farmers with whom you once intermarried and traded, where the livestock trample crops and sow discord.

For decades the world has blamed your plight on your own mismanagement of natural resources, and now you have been accused of genocide by the most powerful government on the planet. But then you discover proof positive—as far as science can provide it—that the rain no longer falls because the richest and most powerful nations have been polluting our great aerial ocean, and in so doing have ground the Sahel's people to dust under the heel of famine, poverty and conflict. What is the price of this injustice?

Allow the issue to ramify to the Arctic, the politically influential farmers of Australia, the inhabitants of the world's coastal resorts and the rest of the world, and you will see that climate change could spawn a whole new industry of litigation against those who polluted knowingly and without concern.

The first drops of the deluge to come are already falling, and nowhere are they dropping as fast as in that litigator's paradise the United States. In July 2003 three New England states announced that they would sue the federal government, and by October ten northeastern states had joined forces to sue the Federal Environment Protection Authority to force regulation of CO_2 as a pollutant. (It was a timely act given coal lobbyist and Cheney confidant Quin Shea's 2001 boast that, 'We're taking steps right now to reverse every piece of paper that the EPA has put together where they could call CO_2 a pollutant.'[110]) Just where this challenge will end is uncertain, yet even before it has been decided, other compelling briefs have surfaced.

It should not be too difficult to apportion blame for climate disasters in a court of law, for it is possible to estimate how many extra gigatonnes of CO_2 there are in the atmosphere, for example, as a result of the Global

Climate Coalition's activities. And from that it is possible to calculate how much they have contributed to warming the planet. That warmth translates into a climatic impact to which a dollar figure can be attached. Given the legal wranglings in which the tobacco and asbestos industries have been embroiled, it is easy to imagine that past members of the Global Climate Coalition may become swept up in similar lawsuits.

An interesting legal challenge came late in 2004, when the Inuit sought a ruling from the Inter-American Commission on Human Rights concerning damages wrought by global warming to the culture of the 155,000-strong group. These damages result from a rate of climate change twice that of the global average. Not only is their traditional food—seals, bear and caribou—vanishing, but their land, in some instances, is disappearing under their feet. The Alaskan village of Shishmaref is becoming uninhabitable due to rising temperatures that are reducing sea ice and thawing permafrost, making the shoreline vulnerable to erosion.[156] Hundreds of square metres of land and over a dozen houses have already been lost to the sea, and there are plans to relocate the whole town—at a cost of over $100,000 per resident.[170]

Shishmaref's plight is particularly poignant. Its population is only 600 strong, but it has persisted at least 4000 years, and its inhabitants look set to become the first climate change refugees. Where they will go remains uncertain, for as they see it:

> the Arctic is becoming an environment at risk in the sense that sea ice is less stable, unusual weather patterns are occurring, vegetation cover is changing, and particular animals are no longer found in traditional hunting areas during specific seasons. Local landscapes, seascapes, and icescapes are becoming unfamiliar, making people feel like strangers in their own land.[156]

Although the commission to which the Inuit have appealed has no legal powers, a favourable ruling may enable them either to sue the US government in an international court, or US corporations in a federal court. In either case it is likely that the Inuit will refer to both the

Universal Declaration on Human Rights, which states that 'everyone has the right to a nationality' and that 'no one shall be arbitrarily deprived of his property', and to the United Nations Covenant on Civil and Political Rights, which states 'that in no case may a people be deprived of its own means of subsistence'. Ultimately the case may be about far more than that, for so immense are changes in the Arctic that the Inuit may be the first people to see their nation—the land and the way of life it supports—extinguished.

The death of a nation has extraordinary implications, as anthropologist Jon Barnett of the University of Melbourne and his colleague Neil Adger point out: 'For all states to do less than everything possible to prevent the loss of sovereign entity is to undermine [the] most essential and powerful norm of international law and politics.'[157] There is no term as far as I know for the extinction of a sovereign state. Perhaps we will soon need to invent one.

Other inhabitants of lands immediately vulnerable to climate change are those of the five sovereign atoll countries. Atolls are rings of coral reef that surround a lagoon, and scattered around the reef crest are islands and islets, whose average height above sea level is a mere two metres. Kiribati, Maldives, Marshall Islands, Tokelau and Tuvalu—which between them support around half a million people—consist only of atolls.

As a result of the destruction of the world's coral reefs, rising seas and the intensifying weather events already in train, it seems inevitable that these nations will be destroyed by climate change during the course of this century. Given the precariousness of their position, you might be surprised by their lack of action on climate change in international forums. This hasn't been due to laziness, but the result of bullying by one of the worst CO_2-emitting nations—Australia.

Political negotiations are often brutal, but the lead-up to Kyoto saw Australia behaving in particularly distasteful ways. Most reprehensible was the coercion of its Pacific Island neighbours into dropping their stance that the world should take 'firm measures' to combat climate change. 'Being small, we depend on them so much we had to give in,' said Tuvalu's Prime Minister Bikenibu Paeniu, following the South Pacific Congress in which Australia laid its demands on the table.

In what must be one of the most outrageous comments uttered in this context, the Australian government's chief economic adviser on climate change, Dr Brian Fisher, told a London conference that it would be 'more efficient' to evacuate small Pacific Island states than to require Australian industries to reduce their emissions of carbon dioxide.[118] With this chilling arrogance ringing in their ears, the Tuvaluans took the only course open to them: they negotiated migration rights to New Zealand for the entire population in the event of serious climate change impacts.[157]

Even where nations are not so threatened by climate change, there will still be big winners and big losers. Under existing projections, just two countries—Canada and Russia—will reap 90 per cent of the benefit that global warming brings to food crops, while other regions such as Africa and India will lose out heavily with only a small degree of warming.[188] Even conservative studies predict a tripling in the number of humans at risk of food shortage by the 2080s, and such changes may bring issues of natural justice to centre stage in our thinking.[188] Nor will health issues be immune. As our globe warms a degree or two, the percentage of humans exposed to malarial parasites will rise from 45 to 60 per cent.[188] What of those people living in regions marginal for malaria today, who are almost certain to be affected? Add to this rising seas, changing storm tracks, rainfall and heatwaves, and you get a sense of the scope of legal action possible in a world without Acts of God. Perhaps, in future, an international court will be created and asked to arbitrate on such matters.

With all of this in mind, it's hard to avoid the idea that any solution to the climate change crisis must be based upon principles of natural justice. After all, if democratic governments do not act voluntarily according to these principles, the courts are liable to force them to do so.

In that case the principle of 'polluter pays' will become paramount, for this principle also implies that the polluter should compensate the victim.

Prior to the Kyoto Protocol all individuals possessed an unfettered right to pollute the atmosphere with greenhouse gases. Now, only the ratifying nations have an internationally recognised right to pollute within limits. Where, one wonders, does this leave the non-ratifying signatories? It is a matter that must be under consideration in chambers of law around the world.

THIRTY-THREE

2084: THE CARBON DICTATORSHIP?

> If…man encroached upon Gaia's functional power to such an extent that he disabled her[,] he would wake up one day to find that he had the permanent, lifelong job of planetary maintenance engineer…Then at last we should be riding that strange contraption, the 'spaceship Earth', and whatever tamed and domesticated biosphere remained would indeed be our 'life support system'.
>
> James Lovelock, *Gaia,* 1979.

Paul Crutzen helped save the world from ozone depletion by CFCs, and for it he received a Nobel prize. With the threat of climate change looming Crutzen is once again engaging in the debate, and he is already thinking far ahead. 'Our future may well involve internationally accepted, large-scale geo-engineering projects…to optimise climate,' he opined in *Nature* in 2002. It is a thought that merits time to explore, and to begin we must look at the great game of climate modification that humanity is engaged in.

I can see three possible outcomes:

1) Our response to limiting emissions is too slow or unco-ordinated to avert great climate shifts, which destroy Earth's life support systems and destabilise our global civilisation. As a result humans are thrust into a protracted Dark Ages far more mordant than any that has gone before, for the most destructive weapons ever devised will still exist, while the means to regulate their use, and to make peace, will have been swept away. These changes could commence as soon as 2050.

2) Humanity acts promptly—on individual, national and corporate levels—to reduce emissions, and so avoids serious climatic consequences. Based on current trends, we will need to have commenced significant decarbonising of our electricity grids by around 2030, and to have substantially decarbonised transport systems by 2050. If we are successful, by 2150 or thereabouts greenhouse gas levels will have dropped to the point where Gaia can once again control Earth's thermostat.

3) Emissions are reduced sufficiently to avoid outright disaster, but serious damage to Earth's ecosystems results. With world climate on a knife-edge, Crutzen's vision of internationally agreed geoengineering projects becomes mandatory. Civilisation will hover on the brink for decades or centuries, during which period the carbon cycle will need to be strictly controlled, by large and small geoengineering projects alike.

Under this final scenario humans would have no choice but to establish an Earth Commission for Thermostatic Control, something that could easily grow from the Kyoto Protocol. Let's begin by considering how the commission might deal with CO_2—the most significant of the thirty-odd greenhouse gases it would need to concern itself with. Among its most important early tasks—and one already of concern to Kyoto— would be to maintain the value of the carbon dollar by arbitrating wherever carbon trade deals are not honoured, and where sequestered carbon is lost. Because of the long time scales involved in projects such as forestry plantations and sequestration proposals, the commission would find itself monitoring carbon dollars minted in 2005 for centuries to come.[173]

It is likely that the commission would need to use the oceans as a tool to regulate the Earth's thermostat. This will require new international co-operation on the use and ownership of the global oceanic commons, and it is possible that the Arctic and Antarctic will also become enmeshed in these new agreements, thus regulating our last significant global commons. Because of the importance of soils as carbon sinks, the commission would be deeply interested in agriculture and land use worldwide and we could anticipate far-reaching regulations dealing with agriculture, forestry and other land uses.

As the climate crisis deepens the commission may also be called upon to arbitrate in circumstances where one nation is suffering gross disadvantage as a result of changed climate, while others are prospering. Australia, for example, may find itself on the brink of collapse as a result of declining rainfall across its main population and agricultural centres, while Canada may enjoy bonanza harvests and mild winters because of the very same climate changes.

If such a commission were to take root, its powers and influence would widen and deepen with the climate crisis, and eventually it would, of necessity, impinge on what hitherto have been sovereign matters.

It is difficult to imagine that such moves, no matter how necessary they might be to stabilise global climate, would not be challenged by some nations. One could expect both foot-dragging and deceit, but outright refusal to comply may also be possible. How would the commission deal, for instance, with 'free riders' who ignore its regulations to the detriment of all?

Countries standing behind the commission would seek an array of incentives, including sanctions, which have in the past proved indispensable in ensuring that no nation gets a 'free ride' on the back of any international treaty. And for these punitive measures to have maximum effect, an international court would be required, and—in cases of last resort—an international armed force for use against recalcitrants. Perhaps they will wear green helmets rather than blue, but the UN peacekeepers offer a fine model as to how this arm of the commission might first evolve.

So delicate is our atmosphere, and so vast is the human burden now being placed upon it, that the work of our commission could not possibly cease with the greenhouse gases: even the hydrogen economy may come under its purview. Molecular hydrogen is a trace atmospheric gas which is present at just a half a part per million in the atmosphere, with a life span of just two years. The future hydrogen economy requires the annual transport of several times the total amount of hydrogen present in the atmosphere today and, as we have seen, hydrogen is liable to leak.[178] By replacing half of current fossil fuel use with hydrogen we risk doubling its atmospheric concentration.

One of hydrogen's most significant unwanted properties is its capacity to increase methane abundance by up to 4 per cent. As the gas economy is viewed as a transition to the hydrogen economy, this may have severe greenhouse consequences for a world already overburdened with fugitive methane emissions. Furthermore, the main sink for atmospheric molecular nitrogen is the nitrogen-fixing micro-organisms in the soil, and the consequences of increasing molecular hydrogen there is unknown.[71] There is even the possibility that, when used on a scale large enough to power the world's transport fleet, hydrogen may affect stratospheric water vapour, planetary temperature and ozone. As the leading researchers in the field recently pointed out, 'evaluation of the climate impacts of a [hydrogen economy] have just begun'.[177]

As the tinkering of chemists becomes ever more sophisticated, and as our awareness of atmospheric impacts grows, we can expect that more and more planetary processes will interest our commission. And with so many problems to confront, some commissioners might begin to feel like the boy who plugged the dyke with his finger, only to find the leaks breaking out all around him. They will surely realise that, while the human population remains so large, the stream of new issues threatening climate security will be endless.

Inevitably, one day some commissioner will suggest that their work would be more effectively done were they to concentrate on the root cause of the issue—the total number of people on the planet. And with such a move the Earth Commission for Thermostatic Control will have transformed itself into an Orwellian-style world government with its own currency, army and control over every person and every inch of our planet. As horrific as such an outcome is, if we delay action to combat the climate crisis, the carbon dictatorship may become essential for our survival.

Not 250 years ago, wild lads from the Scottish highlands who knew nothing of the English language, money or trousers, drove the cattle that were their only wealth to English market towns so they could purchase a few luxuries such as gunpowder and salt. Today no citizens of any developed nation has such mastery over their lives as was possessed by those wild highlanders, for we are the descendants of those who swapped such 'freedom' for stable government, three square meals a day, easy transport, and the sophisticated machines that tell us about climate change.

And at times we have relinquished further freedoms in order to confront dire threats. The Founding Fathers of the United States of America created the greatest nation the world has ever seen, and they did so because they feared an overwhelming external threat—the British Crown. Creating the United States was not easy, for the gentlemen of the south, who loved horse-racing, the theatre and their slave plantations, had to reconcile themselves with Puritan New Englanders, who held such things to be the work of the devil. Yet somehow the deal was brokered, and with it every one of the thirteen signatory states ceded a significant portion of their sovereignty. The Founding Fathers created—

with great success—a political entity of sufficient mass to meet the challenges at hand, yet with sufficient safeguards to allow liberty to flourish.

Humans have come such a very long way in such a brief time that our imaginations are irretrievably mired in the past. Perhaps, as seems to be the case with many American neo-conservatives, they are trapped on the western frontier or in the last great war. For others they reside in now obsolete national identities or ideologies. And because our imaginations still dwell in these vanished landscapes, our response to the threat of climate change can seem nonsensical. It is this fact, I think, that has prompted some conservatives to ignore the threat of climate change, while at the same time so jealously protecting our 'freedom'.

If big coal, big oil and their allied interests continue to prevent the world from taking action to combat climate change, we may soon have an Earth Commission for Thermostatic Control. The only way to avoid both tyranny and destruction is to act as America's Founding Fathers did, by swiftly heeding the call to action and by ceding just enough power to a higher authority to combat the threat. And this will only be effective if we act now, before the crisis becomes full blown.

THIRTY-FOUR

TIME'S UP

It has been the consideration of our wonderful atmosphere in its various relations to human life, and to all life, which has compelled me to this cry for the children and for outraged humanity...Let everything give way to this...Vote for no one who says 'it can't be done'. Vote only for those who declare 'It shall be done'.

Alfred Russel Wallace, *Man's Place in the Universe*, 1903.

If everyone who has the means to do so takes concerted action to rid atmospheric carbon emissions from their lives, I believe we can stabilise and then save the cryosphere. We could save around nine out of every ten species currently under threat, limit the extent of extreme weather events so that losses of both human life and investments are a fraction of those being predicted and reduce, almost to zero, the possibility of any of the three great disasters occurring this century.

But for that to happen, individuals, industry and governments need to act on climate change now: the delay of even a decade is far too much. Credible data indicates that the world may experience the end of cheap oil sometime between now and 2010.[187] The few years we have left before the onset of the oil shortage are the crucial ones for making the transition to a carbon-free economy, for that is when we can build the new infrastructure and technologies most easily and at the least expense.

The people in the hot seat today are the CEOs of large energy corporations. Some seem to be hoping that climate change will just disappear, at least until their retirement. The worst are aggressively pushing for more coal-fired power plants, and their influence should not be underestimated: even in New South Wales, whose premier is a well-known environmentalist and which is suffering the worst drought on record, they look set to build new coal-fired power stations. And this despite the fact that the existing power plants consume a fifth as much water as Sydney's 4 million residents!

Whatever their views on climate change, all energy company CEOs have a few things in common. All have responsibilities to a board, shareholders and their employees, and you can be sure that they are fully briefed on the emerging disaster; they can make no plea of ignorance. Furthermore, market reform in the energy sector means that all are increasingly vulnerable to the mood of the market, which is why the actions of consumers and investors are so important.

The dilemma facing the coal burners is a difficult one, yet it is not insoluble. Just as big oil is getting into gas, so should big coal be getting into something else. This may seem a hard line to argue when coal prices stand at an all-time high—but that is what the oil companies have done and are doing, and for very much the same reasons: limitations of supply or pollution sinks dictate that neither oil nor coal has a long-term future. So what could big coal do to move on?

Biomass (fuel derived from crop waste or other plant matter) is just young coal, so it seems a natural step for big coal is to invest in this

emerging technology. Global dimming indicates that we will be required
to take CO_2 out of the atmosphere in order to stabilise Earth's climate.
This could be achieved by burning biomass and sequestering the CO_2
generated, would go some way towards undoing the damage done by
the industry in the past. Coal miners would need assistance to make the
transition to biomass, and governments could help by mandating that a
certain percentage of all fuel burned are biofuels.

But would industry really walk away from all that coal in mines and
undeveloped reserves? Arthur C. Clarke realised that Earth's coal
reserves represent a potent tool in the planetary maintenance engineer's
toolkit. He knew that Milankovich's cycles have not vanished and,
provided Earth's climate is not bumped into a new ultra-hot state, within
a few thousand years our planet will face a cooling that will herald a new
ice age. What is humanity to do then?

If the governments of the world had forbidden further exploration
for new coal reserves, and had purchased all existing stock, the coal that
today is our enemy might become a powerful tool for protecting ourselves
from the onset of the ice. The Arthur C. Clarke Fund for Avoidance of
the New Ice Age could be incorporated into the Kyoto Protocol, and the
nations of the world could contribute to the coal purchases proportion-
ately to their capacity to do so.

There are many other things that governments could do to assist both
the consumer and industry in their efforts, both locally and globally. The
most important is to ban the building or expansion of old-fashioned coal-
fired power plants. This would send a strong signal to the market as to
future directions of energy production. Good energy efficiency legislation
is equally important, and should be part of every government's thinking.
That includes ever-stricter regulation on the efficiency of goods allowed
into the marketplace, strong housing codes that mandate a limit to
emissions at the household level, legislation that encourages the retro-
fitting of devices that reduce household emissions, and designing transport
systems with overall energy efficiency in mind. It is also important that

cross-subsidies be removed—big energy users like smelters will never feel the full impact of price signals (and thus will never get serious about efficiency) while we householders are footing the bill for much of their power use.

Initiatives to encourage the use of renewable energy are equally important, and could include telling energy providers that they must source a percentage of their energy from renewables (called mandated renewable energy target schemes); rebates for the purchase of photo-voltaic cells; assistance with the location of electricity inter-connectors that favour renewable sources; and legislation that facilitates the intro-duction of renewables such as wind. This is a mere sampling of what can be done, and it is likely that your government is already doing one or more of these things. (For a more comprehensive list, see the actions listed by the International Climate Change Task Force.[217])

Looking further ahead, there is a democratic, transparent and simple form of international agreement that might one day replace Kyoto. Known as Contraction and Convergence (C&C), it has been championed by UK politician Aubrey Meyer for over a decade.

In some ways C&C is an ultra-democratic variant of the Kyoto Protocol, for at its heart is the simple idea that the only equitable way to reduce emissions is to grant every human being an equal 'right to pollute' with greenhouse gases. As with Kyoto, this right could be traded, though under C&C the volume of trade is likely to be far larger than under Kyoto. In order to understand why, let us look at Americans as an example.

Americans emit three times more CO_2 per person per year, than Europeans, and over a hundred times more than the citizens of the least developed countries. Under C&C, the citizens of the developed countries would need to buy, from the world's poor, sufficient carbon credits to cover their emissions. The trade would take place on a country-to-country basis (rather than individual-to-individual) and would represent a massive wealth transfer. The spur to reduce emissions that this represents is enormous, and this is the 'convergence' part of the equation, for it will force the CO_2 emissions of all citizens, regardless of wealth, to converge. As the point on which they are converging is far lower than that of today, it also represents a great contraction in total emissions.

In Meyer's view, C&C begins with three steps:

1) Reach an international agreement on a 'cap' on CO_2 concentrations in the atmosphere.
2) Estimate how quickly emissions need to be cut back to reach that target.
3) Estimate the total 'carbon budget' that steps 1 + 2 give us, and divide that budget among the world's population on a per capita basis.[197]

As with Kyoto, this process would necessitate the creation of a carbon currency which Meyer calls the Ebcus, and a pre-distribution of Ebcus, he argues, could be used to fund clean technology and clear international debts.[197] And there is no reason why sometime in the future the Kyoto Protocol could not take up the principal innovations of C&C. Indeed, according to Meyer, a number of parties to the Kyoto agreement approve of the model.

C&C represents a far greater departure from business as usual than does Kyoto. It is strong medicine for a dire malaise, and as with all strong medicine there are potential side effects. One is that the scheme might eventually do away with world poverty and the north–south divide. Not all aspects of the proposal should displease the conservatives, for by including every human being in existence under its umbrella it obliterates concern about 'free riders' in the developing world that exists under Kyoto.

Among its potential downsides is the initial cost to industrialised countries. It is also possible that some developing nations may equate population size with wealth transfer and thus decline to act on family planning programs. No such scheme, however, is without flaws, and this one at least is on the table and has received some support.

Some may see hidden agendas at work in C&C, which raises one great potential pitfall on the road to climatic stability: the propensity for groups to hitch their ideological bandwagon to the push for sustainability. The nuclear lobby is already doing this, but so too is the 'less is more' lobby, who believe that humans must reduce their overall consumption if sustainability is ever to be achieved. Both arguments have their merits, but they derive from an ideological base that has the potential to alienate many people, without whose efforts the climate change war will be lost. When facing a grave emergency, it's best to be single-minded.

There are only two points that still need to be made. The very worst thing for citizens of the developed world to do would be to sit on their hands until something like C&C is adopted. Action is needed now, and the only responsible thing you, as a concerned individual, can do is to reduce your own emissions as far and as quickly as possible.

Finally, government is unlikely to do anything unless people demand it. To stiffen the resolve of your government in respect to climate change, you must put the issue at the top of your agenda when it comes time to vote. As Alfred Russel Wallace said over a century ago, 'Vote for no one who says "it can't be done". Vote only for those who declare "It shall be done".' And don't just ask your politician what their position is. Ask them what they, personally, are doing to reduce their own emissions.

THIRTY-FIVE

OVER TO YOU

Come, then—a still, small whisper in your ear,
He has no hope who never had a fear;
And he that never doubted of his state,
He may perhaps—perhaps he may—too late.
William Cowper, 'Truth'

There is one thing that no CEO can afford to look away from—the melee of buyers and sellers known as the market. It is my firm belief that all the efforts of government and industry will come to naught unless the good citizen and consumer takes the initiative, and in tackling climate change the consumer is in a most fortunate position.

If we were still battling CFCs, the consumer could not generate an alternative product. Indeed, regardless of their vigilance, in the absence of an international agreement like the Montreal Protocol, they would be liable to buy CFCs hidden in things such as motor vehicles and refrigerators. With the CO_2 problem, however, technology can set free almost every household on the planet. In other words, there is no need to wait for government to act. You can do it yourself.

You can, in a few months rather than the fifty years allowed by some governments, easily attain the 70 per cent reduction in emissions required to stabilise the Earth's climate. All it takes are a few changes to your personal life, none of which requires serious sacrifices.

Understanding how you use electricity is the most powerful tool in your armoury, for that allows you to make effective decisions about reducing your personal emissions of CO_2. To begin, pick up and read carefully your electricity bill. Is your bill higher than it was at the same time last year? If so, why? A phone call or email inquiry to your power supplier may help clarify this.

While you are there, ask about a green power option (where the provider guarantees to source a percentage of power from renewables). The green power option can cost as little as a dollar per week, yet is highly effective in reducing emissions. If your provider does not offer a suitable green option, dump them and call a competitor. Changing your power supplier is usually a matter of a single phone call, involving no interruption of supply or inconvenience in billing. If, however, a power monopoly still reigns in your area, you need to lobby the authorities to create a free market. It is possible then, in switching to green power, to reduce your household emissions to zero. All as the result of a single phone call.

If you wish to take more decisive action, the best place for most people to start is with hot water. In the developed world, roughly one third of CO_2 emissions result from domestic power, and one third of a typical domestic power bill is spent on heating water. This is crazy, since the Sun will heat your water for free if you have the right device. An initial outlay is required, but such are the benefits that it is well worth taking out a loan to do so, for in sunny climates like California or southern Europe the payback period is around two or three years, and as the devices usually carry a ten-year guarantee, that means at least seven to eight years of free hot water. Even in cloudy regions such as Germany and Britain you will still get several years' worth of hot water for free.[126]

If you wish to reduce your impact even further, start with the greatest consumers of power, which for most people are air conditioning, heating and refrigeration. If you are thinking of installing any such items you should seek out the most energy-efficient model available. A good rule of thumb is to choose the smallest device to suit your average needs, and consider alternatives: it may be cheaper to install insulation rather than buying and running a larger heater or cooler. It can be difficult to convince children that they need to turn off appliances when they are finished with them. One way to teach them is for a family to examine its power bill together and set a target for reduction. When it's met, give the kids the savings.

I became so outraged at the irresponsibility of the coal burners that I decided to generate my own electricity, which has proved to be one of the most satisfying things I've ever done. For the average householder, solar panels are the best way to do this. Twelve 80-watt panels is the number I granted myself, and the amount of power this generates in Australia is sufficient to run the house. To survive on this quantum, however, our family is vigilant about energy use, and we cook with gas. And I'm fitter than before because I use hand tools rather than the electrical variety to make and fix things. Solar panels have a twenty-five-year guarantee (and often last for up to forty years). With the cost of electricity rising, and because I'll be enjoying the free power they provide well into retirement, I view them as a form of superannuation.

The town of Schoenau in Germany offers a different example of direct action. Some of its residents were so alarmed by the Chernobyl disaster that they decided to do something to reduce their country's dependence on nuclear power. It started with a group of ten parents who gave prizes for energy savings. This proved so successful that it soon bloomed into a citizens' group determined to wrest control of the town's power supply from KWR, the monopoly that supplied them.

They put together their own study, then raised DM 2 million to build their own green power scheme. Eventually they raised over DM 6.5 million—enough to purchase the power supply, grid and all, from KWR—and today the town not only runs its own power supply but a successful consulting business which advises on how to 'green' the grid right across the country. Each year Schoenau's power supply becomes greener, and even the big power users, such as a plastics recycling factory situated in the town, are happy with the result.[221]

It is not feasible right now for most of us to do away with burning fossil fuels for transport, but we can greatly reduce their use. Walking wherever possible is highly effective, as is taking public transport. Hybrid fuel vehicles are twice as fuel-efficient as a standard, similar-sized car, and trading in your four-wheel-drive or SUV for a medium-sized hybrid fuel car cuts your personal transport emissions by 70 per cent in one fell swoop.

For those who cannot or do not wish to drive a hybrid, a good rule is to buy the smallest vehicle capable of doing the job you most often require. You can always rent for the rare occasions you need something larger. A few years from now, if you have invested in solar power, you should be able to purchase a compressed air vehicle. Then, you can truly thumb your nose at all of those power and petrol bills.

Despite the way it often feels, employees wield considerable influence in the workplace. If you want to see your workplace become more green-house aware, ask your employer to have an energy audit done. And remember, if you can cut your emissions by 70 per cent, so can the business you work for. By doing so, in the medium term the business will save both money and the environment. And because society so desperately needs advocates—people who can act and serve as witnesses to what can be done and should be done—by taking such public actions you will be achieving results way beyond their local impact.

As you read through this list of actions to combat climate change, you might be sceptical that such steps can have such a huge impact. But not only is our global climate approaching a tipping point, but our economy is as well, for the energy sector is about to experience what the internet brought to the media—an age wherein previously discrete products are in competition with each other, and with the individual.

If enough of us buy green power, solar panels, solar hot water systems and hybrid vehicles, the cost of these items will plummet. This will encourage the sale of yet more panels and wind generators, and soon the bulk of domestic power will be generated by renewable technologies. This will place sufficient pressure on industry that, when combined with the pressure from Kyoto, will compel energy-hungry enterprises to maximise efficiency and turn to clean power generation. This will make renewables even more affordable. As a result, the developing world—including China and India—will be able to afford clean power rather than filthy coal. With a little help from you, right now, the developing giants of Asia might even avoid the full carbon catastrophe in which we, in the industrialised world, find ourselves so deeply mired.

Much could go wrong with this linked lifeline to climate safety. It may be that the big power users will infiltrate governments further and stymie the renewables sector; or maybe we will act too slowly, and nations such as China and India will have already invested in fossil fuel generation before the price of renewables comes down. Or perhaps the rate of climate change will be discovered to be too great and we will have to draw CO_2 from the atmosphere.

As these challenges suggest, we are the generation fated to live in the most interesting of times, for we are now the weather makers, and the future of biodiversity and civilisation hangs on our actions.

I have done my best to fashion a manual on the use of Earth's thermostat. Now it's over to you.

POSTSCRIPT

As this book was going to press the journal *Science* published proof positive of global warming. A study by James Hansen and colleagues revealed that Earth is now absorbing more energy, an extra 0.85 watts per square metre, than it's radiating to space. That's the amount of heat emitted by two to three miniature light globes (such as those used on Christmas trees) for every square metre of our planet and, as we add CO_2, the amount increases. The energy imbalance is tiny compared to the 235 watts per square metre received from the Sun, but over years and decades it accumulates, and if left long enough it will mean the difference between survival and destruction for our species. Almost wearily, the scientists conclude that their work 'implies the need for anticipatory actions to avoid…climate change'.[229] It's a call that Hansen, a veteran of climate change science and awareness campaigns, has been making for over twenty years. Perhaps now the world will listen.

AFTERWORD

Soon after I completed this book Hurricane Katrina burst upon New Orleans and changed climate history. Then Rita shook Texas, and people began to wonder whether these giant engines of destruction were harbingers of climate change. As I write this, in late September 2005, the Director of the National Hurricane Center in Miami has said that he expects more storms this season.

Anyone looking only at the number of hurricanes that occur in the Americas each year might think that Katrina and Rita are just part of a natural cycle. This is because there are cycles in Atlantic hurricane activity that mask more significant trends. By affecting the Gulf Stream, the Atlantic Multidecadal Oscillation brings variation in hurricane activity each sixty to seventy years.[244] Another cycle alters hurricane activity in the region each decade or so. Both cycles have complex causes relating to ocean currents and the state of the atmosphere. In order to see beyond these cycles and glimpse the immense changes now influencing our weather, we need to understand how hurricanes form, grow and die. Being a category 5 hurricane—the strongest and most destructive there is—Katrina offers a stark example of the full hurricane life cycle.

As with all hurricanes, Katrina started as a mere thunderstorm, in this case in warm waters off the Bahamas. And this embryonic Katrina may have remained a sound and light show had it not been for a particular configuration of atmospheric conditions that helps transform growing thunderstorms into more potent weather events. The first step in this is the development of a tropical storm. These are groups of thunderstorms that circle until they develop a vortex. Very few thunderstorms develop into tropical storms, however, as wind shear usually destroys the vortex, or a turbulent atmosphere or low pressure in the upper troposphere combine to prevent the circulation and build-up of winds. Over the last decade wind shear has been low in the Caribbean, and a high-pressure system has been present in the upper troposphere.

The atmosphere has also been stable. All of these factors have enhanced convection and set the scene for the development of the perfect tropical storm.[245]

It is at this point in the life cycle of a hurricane that a warm ocean becomes really important. Tropical storms intensify into hurricanes only where the surface temperature of the ocean is around 26 degrees Celsius or greater. This is because hot sea water evaporates readily, providing the volume of fuel—water vapour—required to power a hurricane.

Hurricanes are classified on the Saffir-Simpson Hurricane Scale, which runs from 1 to 5. Category 1 hurricanes lack the puff to do real damage to most buildings, but they can generate a 1.5 metre storm surge, flood coastlines and damage poorly constructed infrastructure. Category 3 hurricanes are more dangerous. They generate winds of between 180 and 210 kilometres per hour, and can destroy mobile homes and strip trees of their leaves. Category 5 hurricanes are an entirely different matter. When they make landfall, 250 kilometres per hour winds ensure that there are no trees or shrubs left standing. Nor are there a lot of buildings left. And with storm surges exceeding 5.5 metres that arrive around four hours before the eye of the storm hits, flooding is far more extensive, and routes allowing people to flee are obstructed early.

By the time Katrina slammed into Florida on 25 August, it had developed into a category 1 storm with wind speeds of 120 kilometres per hour. Even so, Katrina killed eleven in Florida. Hurricanes often peter out when they cross onto land, but somehow Katrina survived the transit of the Florida Peninsula and on 27 August emerged into the Gulf of Mexico. During the summer of 2005 the surface waters of the northern Gulf were exceptionally hot—around 30 degrees Celsius. This, incidentally, is too warm to enjoy a swim. Large bodies of sea water don't get much hotter, and the Gulf waters are deep, which provides a large heat reservoir. Such waters yield vast volumes of water vapour, and during its four-day passage through Gulf waters Katrina grew and grew, until it reached category 5.

By the time Katrina neared New Orleans it had subsided into a category 4 storm, the eye of which passed fifty kilometres east of the city. Thus, Katrina was not the fiercest of storms when it struck, nor did it score a direct hit on the city. Yet its impact was catastrophic. Half a million people inhabited the city, large parts of which lay several metres below sea level—a key factor in its vulnerability. The levees that kept the waters of the Mississippi and Lake Ponchartrain at bay had been built with a kinder climate in mind, and could not withstand the impact of a category 4 or 5 hurricane. With the number of very powerful hurricanes increasing over the past decade it was widely understood that the devastation of the city was only a matter of time. In October 2004 *National Geographic* ran a story outlining the dangers, and in September 2005 *Time* again listed what they were.

So many things went wrong in New Orleans. Poverty, high levels of gun ownership, and official corruption and incompetence, all combined to stymie the relief effort. And then there was the industrial pollution released by the storm surge and high winds. In a region that supplies and refines a considerable proportion of America's oil, spills were inevitable. While estimates of the volumes of pollutants released are not yet available, they must be considerable, for Katrina flooded many of the 140 large petrochemical works that comprise Louisiana's 'cancer corridor'. This damage, of course, was magnified by Rita, which hit at the heart of the US petrochemical industry in Texas.

All of this teaches us that many of the most devastating impacts of any individual hurricane are not related to global warming. Whether Katrina was a little weaker or stronger, whether she struck fifty or 150 kilometres from the city, and whether she struck a week earlier or later, are all matters of chance. But equally there is growing evidence that global warming is changing the conditions in the atmosphere and oceans in ways that will make hurricanes even more destructive in future.

Let us first look at how global warming may be influencing the formation of hurricanes. The Gulf Stream is an important factor here, and there is clear evidence that global warming is affecting its speed. Whether this change will lead to more hurricane activity—or less—is as yet unclear: but conditions are indisputably changing. The state of the upper troposphere is also important, and this is influenced by the tropopause (where the troposphere and stratosphere meet, see p. 21). Both ozone depletion and greenhouse gas accumulation are changing the energetics of the tropopause in ways that can affect hurricane formation. Much more research is needed before the significance of these changes is fully understood, but the very fact of their existence gives climatologists cause for concern.

The impact of climate change on the later phases of the hurricane life cycle is more certain. Satellite measurements reveal that the oceans are rapidly warming from the top down as the result of additional heat coming from the atmosphere. Already the oceans have warmed on average by half a degree Celsius, though some areas—such as the Gulf of Mexico—have warmed far more. In response to this, the amount of water vapour (hurricane fuel) in the air over the oceans has increased by 1.3 per cent per decade since 1988.[245] Both the warmer ocean and the increased water vapour increase the energy available for all manner of storms, from thunderstorm to hurricanes. But it is especially important in transforming tropical storms into hurricanes, and in feeding category 1 hurricanes so that they become category fives. With this enhancement of hurricane fuel, Katrina was an accident just waiting to happen.

The link between warm sea water and hurricane activity was recently strengthened when geologists coring in the Gulf of Carpentaria, between Australia and Papua New Guinea, encountered finely laminated sediments that were laid down in a huge lake during the ice age, a time when sea surface temperature was a few degrees cooler than now.[246] There are many similarities between the Gulf of Carpentaria and the Gulf of Mexico. Both regions are notorious for hurricanes today, so it surprised the scientists to discover that the fine layers bear no evidence of disturbance by storm surges or great waves. This indicates that

Australia's worst hurricane region was, when the ocean was a little cooler, untroubled by great storms for thousands of years.

So is the recent warming of the oceans responsible for the increase in hurricane activity seen in recent years? In September 2004 Dr Thomas Knutson of the National Oceanic and Atmospheric Administration (NOAA) and Dr Robert Tuleya of the Centre for Coastal Physical Oceanography in Norfolk, Virginia, published a comprehensive computer study demonstrating how hurricanes should react to increasing levels of CO_2 in the atmosphere (and thus increasing sea temperatures).[247] The computer models assumed that CO_2 would reach 760 parts per million (around twice current levels) by the 2080s. This change yielded a 14 per cent increase in the intensity of the average hurricane, an increase of maximum surface wind speeds of 8 per cent, and an increase in precipitation (within 100 kilometres of the storm centre) of 18 per cent. Big changes such as these are capable of doing considerable damage to infrastructure.

What is increasingly perplexing and astonishing meteorologists is that, in the real world, we are already seeing an increase in hurricane intensity and numbers far in advance of that suggested by the computer modelling. Dr Kerry Emanuel of the Massachusetts Institute of Technology has found that the total amount of energy released by hurricanes worldwide has increased by 60 per cent in the last two decades.[248] And Dr Peter Webster of the Georgia Institute of Technology in Atlanta (Science) has discovered that more of that energy is going into the most powerful hurricanes. Since 1974 the number of category 4 and 5 hurricanes recorded has almost doubled.[249]

Some commentators believe the discrepancy between the computer models and conditions in the real world somehow indicate that global warming is not responsible for the increasing cyclone activity. Others, however, believe it suggests what they have long suspected: that the global circulation models used to simulate future changes in climate are deeply conservative. If these researches are correct, then the current heat imbalance of the Earth has been sufficient to shift our planet's climate into a new, more dangerous phase.

Much hangs upon this scientific debate. When Hurricane Ivan roared through the Gulf of Mexico in 2004, it disrupted oil production with the tallest waves ever recorded in the region. They tore up long stretches of underwater pipeline, inflicting much more damage than occurred on the surface. The oil industry considered Ivan to be a one in 2500-year event, but then came Katrina and Rita. 'We're seeing one hundred year events happening every few years', one oil industry executive said.[250] Adapting to these changes will be costly, and the investment will only be made if it is clearly warranted.

Cities are manifestations of climate, for they depend upon the services that climate delivers—including a stable sea level, sufficient rainfall, and protection from extreme weather events. Hurricanes like Katrina can change the topography of our planet in ways that leave cities more vulnerable to the next big hit. The Chandeleur Islands used to protect the Mississippi delta from the open waters of the Gulf of Mexico, because the seventy-kilometre-long barrier off the Louisiana Coast would damp down storm surges and waves. When Dr Lawrence Rouse of the Louisiana State University searched for the islands in the wake of Katrina, he discovered that they'd 'pretty well gone'.[251] At the same time the entire delta is sinking into the ocean, making it ever more vulnerable to extreme weather events.

President George Bush has pledged to rebuild New Orleans. The costs will be huge, and the wisdom of making this investment depends very much upon whether the underlying conditions that give rise to powerful hurricanes have indeed shifted. Getting good advice on this, from the American scientific community at least, is not going to be easy, for the Bush administration has made it clear that they want to hear nothing of climate change from the scientists in their employ.

Relationships between the scientists and the administration recently struck a new low. Senator Joe Barton of Texas chairs the powerful House

Energy and Commerce Committee, and is one of the oil lobby's best friends. In June 2005 he used his position to strike out at three of the country's most eminent climate researchers, including Professor Michael Mann at the University of Virginia, co-author of the so-called hockey stick graph which shows how Earth's temperature has varied over the past millennium. According to the *Washington Post*, Barton wrote demanding information about what he claimed were "methodological flaws and data errors" in their studies of global warming. Barton's letters to the scientists had a peremptory, when-did-you-stop-beating-your-wife tone. Mann was told that within less than three weeks, he must list "all financial support you have received related to your research", provide "the location of all data archives relating to each published study for which you were an author", "provide all agreements relating to . . . under-lying grants or funding" and deliver similarly detailed information in five other categories.'[252]

Even other Republicans were shocked by this crude bullying. Republican Sherwood Boehlert of New York, for example, wrote to Barton stating that the purpose of his investigation appeared to be to 'intimidate scientists rather than to learn from them, and to substitute Congressional political review for scientific peer review'.[252] Powerful people have often shot the messenger, but with so much at stake here, America would be well served if its scientists felt able to offer frank and fearless advice.

Despite the current surge in hurricane activity, the full impact of climate change as predicted by the computer models may still be decades away. Yet if we carry on burning fossil fuels as we are now, they are probably inevitable. It's even possible that in the new climate, vast engines of destruction will visit cities as far afield as Washington, New York, Brisbane and Sydney.

Hurricanes have such a catastrophic impact that they focus attention on climate change in a way that few other natural phenomena do. And they have the potential to kill many more people than the largest terrorist attack. Living with a heightened risk of such devastation should act as a constant reminder to us that the failure to combat climate change carries a high price indeed.

CLIMATE CHANGE CHECKLIST

ACTION		IMPACT
Change to an accredited Green Power option	=	Eliminate household emissions from electricity
Install solar hot water system	=	Up to 30% reduction in household emissions
Install solar panels	=	Eliminate household emissions from electricity
Use energy-efficient whitegoods	=	Up to 50% reduction in household emissions from electricity
Use a triple-A rated shower-head	=	Up to 12% reduction in household emissions
Use energy-efficient light globes	=	Up to 10% reduction in household emissions
Check fuel efficiency of next car	=	Up to 70% reduction in transport emissions
Walk, cycle or take public transport	=	Can reduce transport emissions
Calculate carbon footprint	=	Can eliminate transport and household emissions
Suggest a workplace audit	=	Up to 30% reduction in emissions
Write to a politician about climate change	=	Can change the world

GREEN POWER

Here is a list of suppliers of Green Power in the UK.

Ecotricity
http://www.ecotricity.co.uk

Green Energy (UK)
http://www.greenenergyuk.com

Good Energy
http://www.good-energy.co.uk

**London Energy –
Green Tariff**
http://www.london-energy.com/
showPage.do?name=home
energy.switchBrand.green.til

**Northern Ireland Electricity –
Eco Energy**
http://www.nie.co.uk/nieenergy/
ecoenergy

npower Juice
http://www.npower.com/
greenelectricity

PowerGen GreenPlan
http://www.powergen.co.uk/pub/
Dom/A/ui/Residential/GreenPlan.
aspx?id=60

**Scottish and Southern Energy
Group – RSPB Energy**
http://www.rspbenergy.co.uk

**Scottish Power & MANWEB –
Green Energy**
http://www.warminside.co.uk

Seeboard – Green Tariff
http://www.seaboard-energy.
com/showPage.do?name=home
energy.switchBrand.green.til

SWEB – Green Tariff
http://www.sweb-energy.com/
showPage.do?name=home
energy.switchBrand.green.til

Utilita
http://www.utilita.co.uk

CARBON FOOTPRINT

You can calculate your carbon footprint, including household, car and air travel emissions and then, for the cost of a coffee, neutralise them. Go to www.climatefriendly.com/calc.php

CAR

For ratings on the environmental performance of new vehicles sold in the UK go to http://www.eta.co.uk

HOUSEHOLD APPLIANCES

For ratings on the most energy-efficient home appliances go to http://www.est.org uk/myhome

For more information please go to http://www.stopclimatechaos.org

ACKNOWLEDGMENTS

This book could not have been written without the help of many people. Alexandra Szalay was the first to read it, and in doing so she improved it immeasurably. Rob Purves actively encouraged me from the start, and a grant from the Purves Foundation for the Environment made it possible for me to complete the work. The Board of the South Australian Museum, Arts SA, Premier Mike Rann and the Minister for Environment and Conservation John Hill each contributed significantly through their willingness to accommodate my extended absences while writing. They have my admiration and thanks.

Without an invitation from Stuart Pimm of Duke University to participate in the First Okazaki Extinction Conference, and the presence there of Steve Schneider (who tuned me in to the scale of the problem) and Steve Williams (who brought home what it all really meant) I never would have started. Nick Rowley, Martin Copley, Graeme Morgan, Vicki Pope and her staff at the Hadley Centre, and Patrick Filmer-Sankey all provided assistance and ideas, and to them I am truly grateful. Many other scientists and social commentators on hearing that I was embarking on this work sent material unsolicited, some of which proved invaluable.

How does one begin to thank those who took time from their busy lives to read a first draft of a book? Jared Diamond, Andrew Stock, Peter Cosier, Clive Hamilton, Dr Eugene FitzPatrick, Allan Pring, Greg Rouse, Greg Bourne, Graham Pearman and Nick Palousis and his team at the Natural Edge Project all provided thoughtful criticism. If *The Weather Makers* helps persuade readers to reduce their CO_2 emissions, it will be due in no small part to the efforts of these experts.

Research assistants David Flannery and Noriko Wynn and helpers Emma Flannery and Naomi Wynn each did an excellent job. Melanie Ostell deserves special thanks for her care in editing what was a challenging manuscript. Finally, Michael Heyward, publisher and friend, provided rare encouragement and patience.

ILLUSTRATIONS

Grateful acknowledgment is made to the following for permission to reproduce illustrative material:

All line art by Tony Fankhauser: p. 129 redrawn from information supplied by the Water Corporation, WA; p. 163 redrawn from information supplied by IPCC.

p. 110 *Gobiodon* Species C: copyright Glenn Barrall.

p. 120 Gastric brooding frog: copyright Michael J. Tyler.

p. 157 Comparison of climate predictor model and actual weather: copyright The Met Office, UK.

REFERENCES

1. Broecker, W. S. 1995. Cooling the Tropics. *Nature* 376, pp. 212–13.

2. Curran, M. A. J. et al. 2003. Ice Core Evidence for Antarctic Sea Ice Decline Since the 1950s. *Science* 302, pp. 1203–06.

3. Nash, J. Madeline 2002. *El Niño: Unlocking the Secrets of the Master Weather-maker.* Warner Books. New York.

4. Pittock, B. (ed.) 2003. *Climate Change: An Australian Guide to the Science and Potential Impacts.* Australian Government Greenhouse Office, Canberra.

5. Wallace, A. R. 1903. *Man's Place in the Universe: A Study of the Results of Scientific Research in Relation to the Unity or Plurality of Worlds.* George Bell & Sons. London.

6. Arrhenius, S. 1896. On the Influence of Carbonic Acid in the Air upon the Temperature of the Ground. *Philosophical Magazine* 41, pp. 237–76.

7. Högbom, A. G. 1894. Om Sannolikheton FöSekulära Forandringar I Atmosfärens Kolsyrehalt. *Svensk Kemisk Tidskrift* 4, pp. 169–77.

8. Neidjie, B. 2002. *Gagadju Man: Kakadu National Park, Northern Territory, Australia.* J. B. Books. South Australia.

9. Fu, Q. et al. 2004. Contribution of Stratospheric Cooling to Satellite-inferred Tropospheric Temperature Trends. *Nature* 429, pp. 55–58.

10. Santer, B. D. et al. 2003. Contributions of Anthropogenic and Natural Forcing to Recent Tropopause Height Changes. *Science* 301, pp. 479–83.

11. Appenzeller, T. 2004. The Case of the Missing Carbon. *National Geographic* 205/2, pp. 88–117.

12. Saleska, S. R. et al. 2003. Carbon in Amazon Forests: Unexpected Seasonal Fluxes and Disturbance-induced Loss. *Science* 302, pp. 1554–57.

13. Wallace, A. R. 1902. *Island Life, or the Phenomena and Causes of Insular Floras, Including a Revision and Attempted Solution of the Problem of Geological Climates.* Macmillan. London.

14. Crutzen, P. J. 2002. The Geology of Mankind. *Nature* 415, p. 23.

15. Crutzen, P. J. & Stoermer, E. F. 2000. The Anthropocene. *IGBP Newsletter* 41, p. 12.

16. EPCA community members 2004. Eight Glacial Cycles from an Antarctic Ice Core. *Nature* 429, pp. 623–28.

17. Richerson, P. J., Boyd, R. & Bettinger, R. L. 2001. Was Agriculture Impossible during the Pleistocene But Mandatory during the Holocene? A Climate Change Hypothesis. *American Antiquity* 663, pp. 387–411.

18. Ruddiman, W. F. 2003. The Anthropogenic Greenhouse Era Began Thousands of Years Ago. *Climatic Change* 61, pp. 261–93.

19. Zachos, J. C. et al. 2003. A Transient Rise in Tropical Sea-surface Temperature during the Paleocene-Eocene Thermal Maximum. *Science* 302, pp. 1551–54.

20. Wells, S. 2002. *The Journey of Man. A Genetic Odyssey*. Princeton University Press. Princeton.

21. Karoly, D. et al. Detection of a Human Influence on North American Climate. *Science* 302, pp. 1200–3.

22. Urban, F. E., Cole, J. E. & Overpeck, J. T. 2000. Influence of Mean Climate Change on Climate Variability from a 155-year Tropical Pacific Coral Record. *Nature* 407, pp. 989–93.

23. Schiermeier, Q. 2004. Modellers Deplore 'Short-termism' on Climate. *Nature* 428, p. 593.

24. Crump, M. 1998. *In Search of the Golden Frog*. University of Chicago Press. Chicago.

25. Kiesecker, J. M., Blaustein, A. R. & Belden, K. 2001. Complex Causes of Amphibian Population Declines. *Nature* 410, pp. 681–84.

26. Laurance, W. F. et al. 2004. Pervasive Alteration of Tree Communities in Undisturbed Amazonian Rainforests. *Nature* 428, pp. 171–75.

27. Parmesan, C. & Yohe, G. 2003. A Globally Coherent Fingerprint of Climate Change Impacts across Natural Systems. *Nature* 421, pp. 37–42.

28. Pouliquen-Young, O. & Newman, P. 1999. *The Implications of Climate Change for Land-based Nature Conservation Strategies*. Final Report 96/1306, Australian Greenhouse Office, Environment Australia, Canberra, and Institute for Sustainability and Technology Policy, Murdoch University, Perth.

29. Pounds, J. A., Fogden, M. L. P., & Campbell, J. H. 1999. Biological Response to Climate Change on a Tropical Mountain. *Nature* 398, pp. 611–15.

30. Richards, S. J., Alford, R. A. & Bradfield, K. S. 2003. The Great Frog Decline in Australasia: Causes, Developments, and Conservation. *Evolution and Biography of Australasian Vertebrates*. J. R. Merrick et al. Australian Scientific Publishers. Sydney.

31. Root, T. L. et al. 2003. Fingerprints of Global Warming on Wild Animals and Plants. *Nature* 421, pp. 57–60.

32. Savage, J. M. 1970. On the Trail of the Golden Frog: With Warszewicz and Gabb in Central America. *Proceedings of the California Academy of Sciences* 38, pp. 273–87.

33. Thomas, C. D. et al. 2004. Extinction Risk from Climate Change. *Nature* 427, pp. 145–48.

34. Visser, M. E. & Holleman, L. J. M. 2001. Warmer Springs Disrupt the Synchrony of Oak and Winter Moth Phrenology. *Proceedings of the Royal Society B.* 268, pp. 289–94.

35. Williams, S. E., Bolitho, E. E. & Fox, S. 2003. Climate Change in Australian Tropical Rainforests: An Impending Environmental Catastrophe. *Proceedings of the Royal Society. B.* 270, pp. 1887–92.

36. Williams, S. E. & Hilbert, D. W. 2005. Climate Change Threats to the Biodiversity of Tropical Rainforests in Australia, from *Emerging Threats to Tropical Forests*. Barlow, J. & Peres, C. (eds). Chicago University Press. Chicago.

37. Berkelmans, R. & Willis, B. L. 1999. Seasonal and Local Spatial Patterns in the Upper Thermal Limits of Corals on the Inshore Great Barrier Reef. *Coral Reefs* 18, pp. 55–60.

38. Done, T. P. et al. 2003. Global Climate Change and Coral Bleaching on the Great Barrier Reef. Final report to the State of Queensland Greenhouse Taskforce through the Department of Natural Resources and Mines. Australian Institute of Marine Science. Townsville.

39. LeClerq, N., Gattuso, J. P. & Jaubert, J. 2002. Primary Production, Respiration, and Calcification of a Coral Reef Mesocosm under Increased Partial CO_2 Pressure. *Limnology & Oceanography* 47, pp. 558–64.

40. Lough, J. M. 2000. 1997–98: Unprecedented Thermal Stress to Coral Reefs? *Geophysical Research Letters* 27, pp. 3901–04.

41. Pockley, P. 2003. Human Activities Threaten Coral Reefs with 'Dire Effects'. *Australasian Science* Jan–Feb. 2003, pp. 29–32.

42. Wilkinson, C. 2002. Coral Bleaching and Mortality: The 1998 Event Four Years Later and Bleaching to 2002. Chapter 1 in *Status of Coral Reefs of the World*. Wilkinson, C. (ed.) Global Coral Reef Monitoring Network. www.aims.gov.au/pages/research/coral-bleaching/scr2002/scr-00.html

43. Gillett, N. P. et al. 2003. Detection of Human Influence in Sea-level Pressure. *Nature* 422, pp. 292–94.

44. Hoerling, M. & Kumar, A. 2003. The Perfect Ocean for Drought. *Science* 299, pp. 691–94.

45. Karoly, D. J. 2003. Ozone and Climate Change. *Science* 302, pp. 236–37.

46. Smith, I. N. et al. 2000. South-west Western Australian Rainfall and Its Association with Indian Ocean Climate Variability. *International Journal of Climatology* 20, pp. 1913–30.

47. Wright, J. & Jones, D. A. 2003. Long-term Rainfall Decline in Southern Australia. *Proceedings National Drought Forum: Science for Drought*, 15–16 April 2003, Brisbane.

48. Williams, V., Noland, R. B. & Toumi, R. 2003. Reducing the Climate Change Impacts of Aviation by Restricting Cruise Altitudes. www.geo-matics.cv.imperial.ac.uk/html/ResearchActivities

49. Hansen, B., Turrell, W. R. & Østerhus, S. 2001. Decreasing Overflow from the Nordic Seas into the Atlantic Ocean through the Faeroe Bank Channel since 1950. *Nature* 411, pp. 927–30.

50. Dickson, B. et al. 2002. Rapid Freshening of the Deep North Atlantic Ocean over the Past Four Decades. *Nature* 416, pp. 832–37.

51. Hyubrechts, P. I. & De Wolde, J. 1999. The Dynamic Response of the Greenland and Antarctic Ice Sheets to Multiple-century Climate Warming. *Journal of Climate* 12, pp. 2169–218.

52. Schar, C. et al. 2004. The role of Increasing Temperature Variability in European Summer Heatwaves. *Nature* 427, pp. 332–36. www.ncbi.nlm.nih.gov/entrez/query.fcgi?

53. Schwartz, P. & Randall, D. 2004. An Abrupt Climate Change Scenario and Its Implications for United States National Security. www.ems.org/climate/pentagon_climatechange.pdf

54. Woodford, J. 2004. Great? Barrier Reef. *Australian Geographic* 76, pp. 37–55.

55. Appenzeller, T. & Dimick, D. R. (with contributions by D. Glick, F. Montaigne & V. Morell). 2004. The Heat Is On. *National Geographic* 206, pp. 2–75.

56. Mastrandrea, M. D. & Schneider, S. 2004. Probabilistic Integrated Assessment of 'Dangerous' Climate Change. *Science* 304, pp. 571–74.

57. Dukes, J. S. 2003. Burning Buried Sunshine: Human Consumption of Ancient Solar Energy. *Climate Change* 61, pp. 31–44.

58. Kump, L. R. 2002. Reducing Uncertainty about Carbon Dioxide As a Climate Driver. *Nature* 419, pp. 188–90.

59. Beerling, D. J. et al. 2002. An Atmospheric CO_2 Reconstruction across the Cretaceous-Tertiary Boundary from Leaf Megafossils. *Proceedings of the National Academy of Science USA* 99, pp. 7844–47.

60. WWF *Living Planet Report 2004*. 9 October 2004.

61. Knoll, A. H. 2004. *Life on a Young Planet*. Princeton University Press. Princeton.

62. Blatt, H. 2004. *America's Environmental Score-card*. MIT Press. Massachusetts.

63. Rind, D. et al. 2004. The Relative Importance of Solar and Anthropogenic Forcing of Climate Change between Maunder Minimum and Present. *Journal of Climate* 17, pp. 906–29.

64. Stouffer, R. J. 2004. Time Scales of Climate Response. *Journal of Climate* 17, pp. 209–17

65. Raymond, P. A. & Cole, J. J. 2003. Increase in the Export of Alkalinity from North America's Largest River. *Science* 301, pp. 88–90.

66. Lackner, K. 2003. Alkalinity Export and Carbon Balance. *Science* 302, p. 985.

67. Mokhov, I. I. et al. 2002. Estimation of Global and Regional Climate Changes during the Nineteenth and Twentieth Centuries on the Basis of the IAP RAS Model with Consideration for Anthropogenic Forcing. *Izvestiya Atmospheric and Oceanic Physics* 38, pp. 555–68.

68. Grant, P. R. 1999. *Ecology and Evolution of Darwin's Finches*. Princeton University Press. Princeton. Rev. edn.

69. Dawkins, R. 2004. *The Ancestor's Tale: A Pilgrimage to the Dawn of Evolution*. Weidenfeld & Nicolson. London.

70. Weart, S. R. 2003. *The Discovery of Global Warming: New Histories of Science, Technology and Medicine*. Harvard University Press. Massachusetts.

71. Schultz, M. G. et al. 2003. Air Pollution and Climate-forcing Impacts of a Global Hydrogen Economy. *Science* 302, pp. 624–27.

72. Thomas, R. et al. 2004. Accelerated Sea-level Rise from West Antarctica. *Science* 306, pp. 255–58.

73. Feely, R. A. et al. 2004. Impact of Anthropogenic CO_2 on $CaCO_3$ System in the Oceans. *Science* 305, pp. 362–66.

74. Sabine, C.L. et al. 2004. The Oceanic Sink for Anthropogenic CO_2. *Science* 305, pp. 367–71.

75. Buessler, K. O. et al. 2004. The Effects of Iron Fertilisation on Carbon Sequestration in the Southern Ocean. *Science* 304, pp. 414–17.

76. Thomas, H. et al. 2004. Enhanced Open Ocean Storage of CO_2 from Shelf-sea Pumping. *Science* 304, pp. 1005–08.

77. Alfred Russel Wallace obituary. 1914. *Geographical Journal* 43, p. 91.

78. Bradley, R. S., Hughes, M. K. & Diaz, H. F. 2003. Climate Change in Medieval Time. *Science* 302, pp. 404–05.

79. Lamb, H. H. 1965. The Early Medieval Warm Epoch and Its Sequel. *Palaeogeography, Palaeoclimatology, Palaeoecology* 1, pp. 13–37.

80. Evelyn, J. 1661. *Fumifugium: Or, the Inconvenience of the Aer, and the Smoake of London Dissipated*. B. White. London. Reprint 1772.

81. Freese, B. 2003. *Coal: A Human History*. Perseus Publishing. Cambridge. Massachusetts.

82. Penner, J. E. et al (eds). 1999. Aviation and the Global Atmosphere. A special report of IPCC working groups 1 and 3, in collaboration with the Scientific Assessment Panel to the Montreal Protocol on substances that deplete the ozone layer. www.grida.no/climate/ipcc/aviation/

83. Yokoyama, Y. et al. 2000. Timing of the Last Glacial Maximum from Observed Sea-level Minima. *Nature* 406, pp. 713–15.

84. Clark, P. U. et al. 2004. Rapid Rise of Sea Level 19,000 Years Ago and Its Global Implications. *Science* 304, pp. 1141–44.

85. Clarke, G. 2003. Superlakes, Megafloods, and Abrupt Climate Change. *Science* 301, pp. 922–23.

86. Roberts, P. 2004. *The End of Oil: On the Edge of a Perilous New World.* Houghton Mifflin & Co. New York.

87. White, M. E. 1993. *The Greening of Gondwana.* Reed Books. Sydney.

88. Manabe, S. & Weatherald, R. T. 1975. The Effects of Doubling CO_2 Concentration on the Climate of a General Circulation Model. *Journal of Atmospheric Science* 32, pp. 3–15.

89. Mitchell, J. F. B. 2004. Can We Believe Predictions of Climate Change? *Quarterly Journal of Meteorological Society* 130, pp. 2341–360.

90. Ohsumi, T., 2004. Introduction: What is Ocean Sequestration of Carbon Dioxide? *Journal of Oceanography* 60, pp. 693–94.

91. Barry, J. P. et al. 2004. Effects of Direct Ocean CO_2 Injection on Deep-sea Meiofauna. *Journal of Oceanography* 60, pp. 759–66.

92. Riebesell, U. 2004. Effects of CO_2 Enrichment on Marine Phytoplankton. *Journal of Oceanography* 60, pp. 719–29.

93. Portner, H. O., Langenbuch, M. & Reipschlager, A. 2004. Biological Impacts of Elevated Ocean CO_2 Concentrations: Lessons from Animal Physiology and Earth History. *Journal of Oceanography* 60, pp. 705–18.

94. Kita, J. & Ohsumi, T. 2004. Perspectives on Biological Research for CO_2 Ocean Sequestration. *Journal of Oceanography* 60, pp. 695–703.

95. Abram, N. J. et al. Coral Reef Death during the 1997 Indian Ocean Dipole Linked to Indonesian Wildfires. *Science* 301, pp. 952–55.

96. Munday, P. L. 2004. Habitat Loss, Resource Specialisation, and Extinction on Coral Reefs. *Global Change Biology* 10, pp. 1642–47.

97. Hughes, T. P. et al. 2003. Climate Change, Human Impacts, and the Resilience of Coral Reefs. *Science* 301, pp. 929–33.

98. Rowan, B. 2004. Thermal Adaptation in Coral Reef Symbionts. *Nature* 430, p. 742.

99. Baker, A. C. et al. 2004. Corals' Adaptive Response to Climate Change. *Nature* 430, p. 741.

100. Ridgwell, A. J., Kennedy, M. J. & Caldiera, K. 2003. Carbonate Deposition, Climate Stability, and Neoproterozoic Ice Ages. *Science* 302, pp. 859–62.

101. Schiermeier, Q. 2004. Global Warming Anomaly May Succumb to Microwave Study. *Nature* 429, p. 7.

102. Trenberth, K., & Hoar, T. J. 1996. The 1990–1995 El Niño-Southern Oscillation Event: Longest on Record. *Geophysical Research Letters* 23, pp. 57–60.

103. Climate Change: Observations and Predictions. Hadley Centre report December 2003. Met Office, UK.

104. Climate Change Scenarios for the United Kingdom. Hadley Centre report April 2002/UKCIP02 Report. Met Office, UK.

105. Ho, C. et al. 2004. Interdecadal Changes in Summertime Typhoon Tracks. *American Meteorological Society* 17, pp. 1767–75.

106. Fyfe, J. 2003. Extratropical Southern Hemisphere Cyclones: Harbingers of Climate Change? *Journal of Climate* 16, pp. 2802–05.

107. Ray, K. C. S. & De, U. S. 2003. Climate Change in India As Evidenced from Instrumental Records. *World Meteorological Organization Bulletin* 52, pp. 53–59.

108. Hollander, J. M. 2003. *The Real Environmental Crisis: Why Poverty, Not Affluence is the Environment's Number One Enemy*. University of California Press. Berkeley.

109. Henschke, I. 1999. Essays on Global Warming and the World Response. Reuters Foundation fellowship paper. Oxford.

110. Kennedy, R. F. 2004. *Crimes against Nature: How George W. Bush and His Corporate Pals Are Plundering the Country and Hijacking Our Democracy*. HarperCollins. New York.

111. Browne, J. 2004. Beyond Kyoto. *Foreign Affairs* July–August. www.foreignaffairs.org/20040701faessay83404/john-browne/beyond-kyoto.html

112. Blair, T. 2004. Speech on climate change to celebrate the tenth anniversary of HRH the Prince of Wales's Business and the Environment Programme, 14 September 2004. www.number-10.gov.uk/output/page6333.asp

113. Koch, P. L., Zachos, J. C. & Dettman, D. L. 1995. Stable Isotope Stratigraphy and Palaeoclimatology in the Palaeocene Bighorn Basin Wyoming, USA. *Palaeogeography, Palaeoclimatology, Palaeoecology* 115, pp. 61–89.

114. Atkinson, A. et al. 2004. Long-term Decline in Krill Stock and Increase in Salps within the Southern Ocean. *Nature* 432, pp. 100–03.

115. Beder, S. The Decline of the Global Coalition. *Engineers Australia* November 2000, p. 41.

116. Lovelock, J. 1979. *Gaia: A New Look at Life on Earth*. Oxford University Press. Oxford.

117. Lutgens, F. K. & Tarbuck, E. J. 2004. *The Atmosphere: An Introduction to Meteorology*. Pearson Prentice Hall. New Jersey. Ninth edn.

118. Hamilton, C. 2001. *Running from the Storm: The Development of Climate Policy in Australia*. UNSW Press. Sydney.

119. Victor, D. G. 2004. *Climate Change: Debating America's Policy Options*. Council on Foreign Relations/Brookings Institute Press. New York.

120. Romm, J. J. 2004. *The Hype about Hydrogen: Fact and Fiction in the Race to Save the Climate*. Island Press. Washington.

121. Liberty Australia. Globaloney Warming. www.alphalink.com.au./ ~noelmed/gwarming.htm

122. Biblebelievers. Myths of Global Warming. www.biblebelievers.org.au/ grreen.htm

123. Schelling, T. C. 2002. What Makes Greenhouse Sense? Time to Rethink the Kyoto Protocol. *Foreign Affairs* May–June.

124. Swing, J. T. 2003. What Future for the Oceans? *Foreign Affairs* September–October, pp. 139–52.

125. McKibbin, W. J. & Wilcoxen, P. J. 2002. Climate Change after Kyoto. *The Brookings Review* 20, pp. 7–10.

126. Girardet, H. 2004. *Cities People Planet: Liveable Cities for a Sustainable World*. Wiley Academy. Chichester.

127. Herman, H. 1952. *Brown Coal*. State Electricity Commission of Victoria. Melbourne.

128. Schiermeier, Q. 2003. The Long Road from Kyoto. *Nature* 426, p. 756.

129. Hawkins, D. G. 2004. Global Warming: Dodging the Silver Bullet. NRDC presentation. www.iea.org/Textbase/work/2004/zets/conference/presentations/hawkins.pdf

130. Coleman, T. 2003. *The Impact of Climate Change on Insurance against*

Catastrophes. www.actuaries.asn.au/PublicSite/pdf/conv03papercoleman.
pdf

131. Europe Goes for Carbon Trading, Sort Of. *Economist*. 1 April 2004.

132. BCSE 2004. Dispelling the Myths about Wind. *EcoGeneration* 24, pp. 8–9.

133. Lucy, S. 2004. Emissions Trading: Is Australia at the Tipping Point? *EcoGeneration* 25, pp. 12–14.

134. Cox, P. M. et al. 2004. Amazonian Forest Dieback under Climate-carbon Cycle Projections for the Twenty-first Century. *Theoretical Applied Climatology* 78, pp. 137–56.

135. Betts, R. A. et al. 2004. The Role of Ecosystem-atmospheric Interactions in Simulated Amazonian Precipitation Decrease and Forest Dieback under Global Climate Warming. *Theoretical Applied Climatology* 78, pp. 157–75.

136. Cox, P. M. et al. 2000. Acceleration of Global Warming Due to Carbon-cycle Feedbacks in a Coupled Climate Model. *Nature* 408, pp. 184–87.

137. Sarkisyan, A. S. 2002. Major Advances and Problems in Modeling Long-Term World Ocean Climate Changes. *Izvestiya, Atmospheric and Ocean Physics* 38, pp. 664–81.

138. Bindschadler, R. A. et al. 2003. Tidally Controlled Stick-slip Discharge of a West Antarctic Ice Stream. *Science* 301, pp. 1087–89.

139. Verburg, P., Hecky, R. E. & Kling, H. 2003. Ecological Consequences of a Century of Warming in Lake Tanganyika. *Science* 301, pp. 505–07.

140. Matamala, R. et al. 2003. Impacts of Fine Root Turnover on Forest NPP and Soil C Sequestration Potential. *Science* 302, pp. 1385–87.

141. Dessai, S. et al. 2004. Defining and Experiencing Dangerous Climate Change. *Climatic Change* 64, pp. 11–25.

142. Bellwood, D. R. et al. 2004. Confronting the Coral Reef Crisis, *Nature* 429, pp. 827–33.

143. Pounds, A. J. 2001. Climate and Amphibian Declines. *Nature* 410, pp. 639–40.

144. Walther, G. R. et al. 2002. Ecological Responses to Recent Climate Change. *Nature* 416, pp. 389–95.

145. Stokstad, E. 2004. Global Survey Documents Puzzling Decline of Amphibians. *Science* 306, p. 391.

146. Comiso, J. C. 2003. Warming Trends in the Arctic from Clear Sky Satellite Observations. *Journal of Climate* 15, pp. 3498–510.

147. Toniazzo, T., Gregory, J. M. & Huybrechts, P. 2004. Climatic Impact of a Greenland Deglaciation and Its Possible Irreversibility. *Journal of Climate* 17, pp. 21–33.

148. Shepherd, A. et al. 2003. Larsen Ice Shelf has Progressively Thinned. *Science* 302:856-858.

149. Curry, R., Dickson, B. & Yashayaev, I. 2003. A Change in the Freshwater Balance of the Atlantic Ocean over the Past Four Decades. *Nature* 426, pp. 826–29.

150. Kerr, R. A. 2004. Sea Change in the Atlantic. *Science* 303, p. 35.

151. Wolff, E. W. 2003. Whither Antarctic Ice? *Science* 302, p. 1164.

152. Kaiser, J. 2003. Warmer Ocean Could Threaten Antarctic Ice Shelves. *Science* 302, p. 759.

153. Rignot, E., Rivera, A. & Cassasa, G. 2003. Contribution of the Patagonia Icefields of South America to Sea Level Rise. *Science* 302, pp. 434–37.

154. Bentley, C. R. 1998. Ice on the Fast Track. *Nature* 394, p. 21.

155. Clarke, T. 2002. Record Melt in Arctic and Greenland. *Nature News* Dec. 2002.

156. Hassol, S. J. 2004. *Impacts of a Warming Arctic: Arctic Climate Impact Assessment*. Cambridge University Press. Cambridge.

157. Barnett, J. & Adger, N. 2003. Climate Dangers and Atoll Countries. *Climatic Change* 61, pp. 321–37.

158. Lovelock, J. 2004. Nuclear Power Is the Only Green Solution. *Independent* 24 May 2004, p. 31.

159. Hughes, L., Cawsley, E. M. & Westoby, M. 1996. Climatic Range Sizes of *Eucalyptus* Species in Relation to Future Climate Change. *Global Ecology and Biogeography Letters* 5, pp. 23–29.

160. Climate Change: Solutions for Australia. 2004. Australian Climate Group. WWF. Sydney.

161. Pacala, S. & Socolow, R. 2004. Stabilization Wedges: Solving the Climate Problem for the Next Fifty Years with Current Technologies. *Science* 305, pp. 968–72.

162. Boodhoo, Y. 2003. Our Future Climate. *World Meteorological Organization Bulletin* 52, pp. 224–28.

163. Bell, J. L., Sloan, L. C. & Snyder, M. 2003. Regional Changes in Extreme Climatic Events: A Future Climate Scenario. *Journal of Climate* 17, pp. 81–87.

164. Stabilisation and Commitment to Future Climate Change. Hadley Centre report October 2002. Met Office, UK.

165. Service, R. F. 2003. As the West Goes Dry. *Science* 303, pp. 1124–27.

166. Zeng, N. 2003. Drought in the Sahel. *Science* 302, pp. 999–1000.

167. Prospero, J. M., & Lamb, P. J. 2003. African Droughts and Dust Transport to the Caribbean: Climate Change Implications. *Science* 302, pp. 1024–27.

168. Gianni, A., Saravanan, R. & Chang, P. 2003. Oceanic Forcing of Sahel Rainfall on Interannual to Interdecadal Timescales. *Science* 302, pp. 1027–30.

169. Karoly, D., Riseby, J. & Reynolds, A. 2003. Global Warming Contributes to Australia's Worst Drought. *Australasian Science* April 2003. www.control.com.au/bi2003/articles243/feat1_243.shtml

170. Whitfield, J. Alaska's Climate: Too Hot to Handle. Nature News Service. 2 October 2003. www.clivar.org/recent/nat_alaska_climate.htm

171. Cook, E. R. et al. 2004. Long-term Aridity Changes in the Western United States. *Science* 306, pp. 1015–18.

172. Kirschbaum, M. U. F. 2003. Can Trees Buy Time? An Assessment of the Role of Vegetation Sinks As Part of the Global Carbon Cycle. *Climatic Change* 57, pp. 227–41.

173. Herzog, H., Caldiera, K. & Reilly, J. 2003. An Issue of Permanence: Assessing the Effectiveness of Temporary Carbon Storage. *Climatic Change* 59, pp. 293–310.

174. Lacour-Gayet, P. 2004. Can the Oil and Gas Industry Help Solve the CO_2 Problem? Plenary address. *APPEA Journal* 2004, pp. 39–46.

175. Canby, T. Y. 1977. The Year the Weather Went Wild. *National Geographic* 152, pp. 799–829.

176. Carbon Dioxide Levels Blow Sky High. Agençe France-Presse. 30 March 2004. www.abc.net.au/science/news/stories/s1076856.htm

177. Prather, M. J. 2003. An Environmental Experiment with H_2? *Science* 302, pp. 581–82.

178. Garman, P. et al. 2003. The Bush Administration and Hydrogen. *Science* 302, pp. 1331–33.

179. Kerr, R. A. 2004. Gas Hydrate Resource: Smaller but Sooner. *Science* 303, pp. 946–47.

180. Keith, D. W. & Garrell, A. E. 2003. Rethinking Hydrogen Cars. *Science* 302, pp. 315–16.

181. Environmental Management of Groundwater: Abstraction from the Gnangara Mound. *Western Australian EPA Bulletin* 1139. Waters and Rivers Commission 2004. www.epa.wa.gov.au/docs/1814_B1139.pdf

182. Siegert, M. J. et al. 2002. The Eurasian Arctic during the Last Ice Age. *American Scientist* 90, pp. 32–39.

183. North Greenland Ice Core Project Members. 2004. High-resolution Record of Northern Hemisphere Climate Extending into the Last Interglacial Period. *Nature* 431, pp. 147–51.

184. Time and Chance. *Economist* 19 December 2003.

185. Deacon, G. Personal correspondence. December 2004.

186. Dickens, G. R. 2004. Hydrocarbon-driven Warming. *Nature* 429, pp. 513–15.

187. Leggett, J. The Coming Crisis. December 2004. www.earthscan.co.uk/news/article/mps/UAN/351/v/3/sp/332685698700328405316

188. Hare, W. 2003. Assessment of Knowledge on Impacts of Climate Change: Contributions to the Specification of Art. 2 of the UNFCCC. WBGU Materialien. Berlin 2003. Also see online.

189. Jones, C. D. et al. 2001. The Carbon Cycle Response to ENSO: A Coupled Climate-carbon Cycle Model Study. *Journal of Climate* 14, pp. 4113–129.

190. Jones, C. D., Cox, P. & Huntingford, C. 2003. Uncertainty in Climate-carbon Cycle Projections Associated with the Sensitivity of Soil Respiration to Temperature. *Tellus* 55b, pp. 642–48.

191. Betts, R. A. 2000. Offset of the Potential Carbon Sink from Boreal Forestation by Decreases in Surface Albedo. *Nature* 408, pp. 187–90.

192. Bush, M. B., Silman, M. R. & Urrego, D. H. 2004. 48,000 Years of

Climate and Forest Change in a Biodiversity Hot Spot. *Science* 303, pp. 827–29.

193. Little, C. T. S. & Vrijenhoek, R. C. 2003. Are Hydrothermal Vent Animals Living Fossils? *Trends in Ecology and Evolution* 18, pp. 582–89.

194. Malthe-Sørenssen, A. et al. 2004. Release of Methane from a Volcanic Basin As a Mechanism for Initial Eocene Global Warming. *Nature* 429, pp. 542–45.

195. Walker, G. 2004. Frozen Time. *Nature* 429, pp. 596–97.

196. Tanser, F. C., Sharp, B. & le Sueur, D. 2003. Potential Effect of Climate Change on Malaria Transmission in Africa. *Royal Society of Tropical Medicine & Hygiene* 97, pp. 129–32.

197. Meyer, A. 2000. *Contraction & Convergence: The Global Solution to Climate Change*. Schumacher Briefing No 5. Green Books for the Schumacher Society. Devon.

198. Parmesan, C. et al. 1999. Poleward Shifts in Geographical Ranges of Butterfly Species Associated with Regional Warming. *Nature* 399, pp. 579–84.

199. Blumberg, M. S. 2002. *Bodyheat: Temperature and Life on Earth*. Harvard University Press. Cambridge.

200. Lopez, B. 1986. *Arctic Dreams: Imagination and Desire in a Northern Landscape*. Scribner & Sons. New York.

201. Cogger, H. G. & Zweifel, R. G. (eds) 1992. *Reptiles and Amphibians*. Southmark Publishers, New York.

202. Judah, T. 2005. The Stakes in Darfur. *New York Review of Books* LII, pp. 12–16.

203. Quarter of City's Water Is out of Reach. *Australian* 2 July 2004.

204. Benton, M. et al. 2003. *When Life Nearly Died: The Greatest Mass Extinction of All Time*. Thames & Hudson. London.

205. Ward, P. 2004. *Gorgon: Palaeontology, Obsession, and the Greatest Catastrophe in Earth's History*. Viking Books. New York.

206. Bond, W. J., Woodward, F. I. & Midgeley, G. F. 2004. The Global Distribution of Ecosystems in a World without Fire. *New Phytologist* 165, pp. 525–38.

207. Hennessey, K. et al. Climate Change in New South Wales. Part 2:

Projected Changes in Climate Extremes. CSIRO November 2004, consultancy report for the NSW Greenhouse Office.

208. Cooper, W. & Cooper, W. T. 1994. *Fruits of the Rainforest*. Geo Productions. Sydney.

209. Goodstein, E. 1997. Polluted Data. *American Prospect* 8, pp. 1–8.

210. Parker, L. 1999. RL30285: Global Climate Change: Lowering Cost Estimates through Emissions Trading—Some Dynamics and Pitfalls. CRS Report for Congress. www.ncseonline.org/nle/crsreports/climate/clim-21.cfm

211. Lash, W. H. 1999. A Current View of the Kyoto Climate Change Treaty. Report for the Center for the Study of American Business. Washington University. St Louis.

212. Energy Innovations: A Prosperous Path to a Clean Environment. Report by Alliance to Save Energy, American Council for an Energy-efficient Economy, Natural Resources Defence Council, Tellus Institute, and the Union of Concerned Scientists. June 1997.

213. Stainforth, D. A. et al. 2005. Uncertainty in Predictions of the Climate Response to Rising Levels of Greenhouse Gases. *Nature* 433, pp. 403–06.

214. Wright, R. 2005. *A Short History of Progress*. Text Publishing. Melbourne.

215. Ainsworth, E. A. & Long, S. P. 2005. What Have We Learned from Fifteen Years of Free-air CO_2 Enrichment FACE? A Meta-analytic Review of the Responses of Photosynthesis, Canopy Properties and Plant Production to Rising CO_2. *New Phytologist* 165, pp. 351–72.

216. Gaia and Selfish Genes: Differing Perspectives on Life. *New York Times* on line: http://endeavor.med.nyu.edu/~strone01/gaia.html

217. Meeting the Climate Change Challenge: Recommendations of the International Climate Change Taskforce. Climate Change Taskforce 2005. Institute for Public Policy Research. London.

218. Fagan, B. 2004. *The Long Summer: How Climate Changed Civilization*. Basic Books. New York.

219. Flannery, T. (ed.) 2002. *Life and Adventures of William Buckley*. Text Publishing. Melbourne.

220. Sloan, E. D. 2003. Fundamental Principles and Applications of Natural

Gas Hydrates. *Nature* 426, pp. 353–59.

221. Suzuki, D. & Dressel, H. 2002. *Good News for a Change: Hope for a Troubled Planet.* Stoddart Publishing. Toronto.

222. Jarrell, J. D., Mayfield, M. & Rappaport, E. N. 2001. The Deadliest, Costliest, and Most Intense United States Hurricanes from 1900 to 2000. *NOAA Technical Memorandum NWS TPC-1.*

223. Diamond, J. 2005. *Collapse: How Societies Choose to Fail or Succeed.* Viking Books. New York.

224. Leggett, J. Dangerous Fiction. Review of Michael Crichton's *State of Fear.New Scientist* 2489, 5 March 2005, p. 50.

225. Przeslawski, R., Davis, A. R. & Benkendorf, K. 2005. Synergistic Effects Associated with Climate Change and the Development of Rocky Shore Molluscs. *Global Change Biology* 11, pp. 1–8.

226. Ending the Energy Stalemate: A Bipartisan Strategy to Meet America's Energy Challenges. The National Commission on Energy Policy 2004. www.energycommission.org/ewebeditpro/items/O82F4682.pdf

227. Feidel, S. 1993. Prehistory of the Americas. Cambridge University Press. Cambridge. Second edn.

228. Schmittner, A. 2005. Decline of the Marine Ecosystem Caused by a Reduction in the Atlantic Overturning Circulation. *Nature* 434, pp. 628–33.

229. Hansen, J. et al. 2005. Earth's Energy Imbalance: Confirmation and Implications. *Science* 308, pp. 1341–45. www.sciencemag.org/cgi/content/abstract/1110252v2?

230. Dotto, L. & Schiff, H. 1978. *The Ozone War.* Doubleday. New York.

231. Overpeck, J., Cole, J. & Bartlein, P. 2005. A 'Palaeoperspective' on Climate Variability and Change. Chapter 7 in Lovejoy, T. E. & Hannah, L. (eds) *Climate Change and Biodiversity.* Yale University Press. New Haven.

232. Hoegh-Guldberg, O. 2005. Climate Change and Marine Ecosystems. Chapter 16 in Lovejoy, T. E. & Hannah, L. (eds) *Climate Change and Biodiversity.* Yale University Press. Newhaven.

233. Sparks, T. H. & Carey, P. H. 1995. The Response of Species to Climate over Two Centuries: An Analysis of the Marsham Phenological Record. *Journal of Ecology* 83, pp. 321–29.

234. Hargroves, K. & Smith, M. 2005. *The Natural Advantage of Nations: Business Opportunities, Innovation and Governance in the Twenty-first Century*. Earthscan. London.

235. Chicago Climate Exchange Reaches Volume Milestone of 1 Million Tons of Carbon Dioxide Traded. Chicago Climate Exchange press release, 1 July 2004. www.chicagoclimatex.com/news/CCXPressRelease_040701.html

236. Braithwaite, J. & Drahos, P. 2000. *Global Business Regulation*. Cambridge University Press. Cambridge.

237. Sustainable Energy Jobs Report: A Report for the Sustainable Energy Development Authority. The Allen Consulting Group 2003. Sydney.

238. Cost Benefit Analysis of New Housing Energy Performance Regulations: Impact of Proposed Regulations Report for Sustainable Energy Authority Victoria and Building Commission. The Allen Consulting Group 2002. Melbourne.

239. Gelbspan, R. 1995. *The Heat Is On: The Climate Crisis, the Cover-up, the Prescription*. Perseus Books. Cambridge.

240. Hamilton, C. & Quiggin, J. 1997. Economic Analysis of Greenhouse Policy: A Layperson's Guide to the Perils of Economic Modelling. Australia Institute. Sydney.

241. Bush Aide Softened Greenhouse Gas Links to Global Warming. *New York Times* 8 June 2005.

242. Wallace, A. R. 1872. *The Malay Archipelago, the Land of the Orang-Utan and the Bird of Paradise: A Narrative of Travel, with Studies of Man and Nature*. Macmillan. London. Fourth edn.

243. And Now for Something Completely Different: An In-depth Response to 'The Death of Environmentalism'. *Grist Magazine* 13 January 2005.

244. Chris Carroll (2005). In Hot Water. *National Geographic Magazine* 208 (2):72—85.

245. Trenberth, K. (2005). Uncertainties in hurricanes and global warming. *Science* 308:1753—1754.

246. De Dekker, P. (2001). Late Quaternary cyclic aridity in tropical Australia. *Palaeogeography, Palaeoclimatology, Palaeoecology* 170:1—9.

247. Knutson, T. R. & Tuleya, R. E. (2004). Impact of CO_2-Induced Warming on Simulated Hurricane Intensity and Precipitation: Sensitivity to the

choice of Climate Model and Convective Parameterization. *Journal of Climate* 17:3477—3495.

248. Emanuel, K. (2005). Increasing destructiveness of tropical cyclones over the past 30 years. *Nature* 436: 686—688.

249. Kerr, R. A. (2005). Is Katrina a harbinger of still more powerful hurricanes? *Science* 309:1807.

250. Jad Mouawad (2005). 110-Year Storm standards for platform criteria may be revised. *International Herald Tribune* 16 September.

251. John Vidal (2005). Tide of Pollution Swamps Industry. *The Age* 10 September.

252. Ignatius, D. (2005). A bid to chill thinking. *Washington Post* July 22.

INDEX